The Learning Garden

PETER LANG
New York • Washington, D.C./Baltimore • Bern
Frankfurt am Main • Berlin • Brussels • Vienna • Oxford

VERONICA GAYLIE

The Learning Garden

Ecology, Teaching, and Transformation

PETER LANG
New York • Washington, D.C./Baltimore • Bern
Frankfurt am Main • Berlin • Brussels • Vienna • Oxford

Library of Congress Cataloging-in-Publication Data

Gaylie, Veronica.
The learning garden: ecology, teaching, and transformation /
Veronica Gaylie.
p. cm.
Includes bibliographical references.
1. School gardens. 2. Ecology–Study and teaching. 3. Outdoor education.
4. Experiential learning. I. Title.
SB55.G34 372.35'7–dc22 2009005535
ISBN 978-1-4331-0470-1

Bibliographic information published by **Die Deutsche Bibliothek.**
Die Deutsche Bibliothek lists this publication in the "Deutsche
Nationalbibliografie"; detailed bibliographic data is available
on the Internet at http://dnb.ddb.de/.

Cover image by Amanda Kehler

The paper in this book meets the guidelines for permanence and durability
of the Committee on Production Guidelines for Book Longevity
of the Council of Library Resources.

© 2009 Peter Lang Publishing, Inc., New York
29 Broadway, 18th floor, New York, NY 10006
www.peterlang.com

Printed in the United States of America

For my students,
past, present and future.

—roam on.

Contents

Acknowledgments

I am grateful to all of the people who lent their support, encouragement and advice to help make this project a reality. First, I thank all of my students, too numerous to mention here, who gave their labour and love to a little piece of land and, somehow, helped make the concept of a 'Learning Garden' real. I especially thank the first students who engaged in the spirit of the garden to develop a space for future learners: Jim, Hadrian, Jess, Dave, Mike M., Mike B., Ryan, Farah and Jenn. Thanks to Mike and Kevin for the fire pit; thanks to Amanda and Bridget for looking after the garden during the hot summer while I wrote. I thank all of my colleagues at UBC Okanagan, who provided support for an idea that was new and different. Special thanks to Robert Campbell. Thanks to Facilities Management, and the wonderful facilities crew. Thanks to the UBC Farm, Vancouver (for surviving). I thank TD Canada Trust Friends of the Environment Foundation and the UBCO Office of the Provost for providing grants. Thanks to Okanagan plant experts Gwen Steele and Simonne Macklem for their knowledge of native plants. I thank all of the small businesses, truck drivers, gravel pit workers, community gardeners, support staff and individuals in the Okanagan who supported the project in so many ways; thank you for your generosity and vision in support of the planet. Special thank-you to my research assistant, (teacher) David Laird.

Many friends, students, colleagues, and mentors, past and present, listened and provided wisdom and encouragement during the journey of writing this book. I thank the Okanagan Nation, the land, air, and water, and the places where the spirit dwells.

I thank Chris Myers and the staff at Peter Lang Publishing in New York.

I am grateful to my parents—who worked hard, provided a solid foundation of care, and constantly encouraged me to dream. I thank my sister, Monica, for the laughter and the listening.

Finally, I owe this book to my husband, Chris, who inspired the idea while walking in the woods of Northern California. *Tha gradh agam ort.*

Thank you to the following individuals for permission to reprint their material:

CHAPTER ONE

❁

The Learning Garden: Introduction

I've been a farmer for over twenty-five years, more than twenty years on the same plot of land...During that time, I've discovered some amazing parallels between agriculture and education. A person who's not a farmer might assume that anyone who had worked on the same piece of land for over twenty years would have it all figured out. Perhaps I am unusually slow, but after all these years I've concluded that instead of having more answers, I actually have more questions. (Ableman 2005, p.177)

Why concern ourselves so much about our beans for seed, and not be concerned at all about a new generation? (Thoreau 1854, p.113)

How does a garden teach? What is the role of environment and community in teacher education? How does learning to teach in the natural world influence how a student approaches the role of 'teacher'? In August 2006 a group of environmental education students and I turned an unused piece of campus land into a 'Learning Garden'. The initial idea was to provide a model garden for schools and a space for interdisciplinary, 'ecology-based' teacher training and exploration of cooperative learning and interdisciplinary, project-based collaboration. Students in the teacher training programme also participated in volunteer community service in order to gain firsthand understanding of the ecological, social and economic circumstances of their local communities. The garden was envisioned as a site where future teachers would learn about environment, even as they learned how to be teachers.

Recently, the media has widely reported scientific evidence of climate change (IPCC 2007; UNEP 2005); the depletion of natural energy sources, the effects of climate change on the world economy (HM Treasury Cabinet Office Environment Policy 2008) and the impact on developing nations (UNEP 2005). Scientific reports also indicate that education and individual and institutional behaviours can aid in mitigating the effects of climate change (IPCC 2007). The facts are startling (Orr 2004):

- At death, human bodies often contain enough toxins and heavy metals to be classified as hazardous waste (p.1);

- US industry releases some 11.4 billion tons of hazardous wastes into the environment each year (p.1);

- Each day, 116 square miles of rain forest are lost; 72 square miles of encroaching deserts increase; 40 to 250 species are lost; population increases by 250,000; 2,700 tons of chlorofluorocarbons and 15 million tons of carbon dioxide are released into the atmosphere (p.7).

In the face of such evidence, is it enough to add an 'ecology' or 'earth' unit of study to a programme plan? Is it enough to change the subject matter of learning or must the processes also be adapted? The way humans use energy in the future alone will dramatically change not only *what* we know, but *how* we live, teach and learn. As well-known environmental educator David Orr explains, "Education has generally prepared the young to live in a high-energy world" (p.58). What, then, is the role of schools in a changing world? In a changing world, how will education lead? Are student teachers ready?

Increasingly, it has become a professional and ethical responsibility to nurture attitudes that place nature at the centre of learning. There is increasingly urgent interest among educators to integrate and develop hands-on learning resources that will engage youth in thought, discussion and action around ecological issues. Perhaps definitions of what it means to be 'educated' need now include a deeper awareness and sense of ethical responsibility for the land in ways that are also local and global, social and ecological knowledge. Such a 'progressive' shift towards tradition, and towards sustainability, will require both new methods and new perspectives in teacher training. Such a shift in thinking towards care, land and community in education will require a deeper transformation; interdisciplinary understanding around a true land ethic will require humility and generosity.

Through research, physical labour and collaborative learning, the Learning Garden on our campus grew as a place where students learned to become teachers and learners. The Learning Garden exposed the new teachers to a concept of the land as both a physical space and an experiential learning process, concepts involving responsible land management, ecological justice, risk taking, community commitment and, ultimately, transformation. But the garden was also a garden. After we pulled weeds, they always returned.

The purpose of this book is to discuss how gardens transform learning through a combination of tangible practice and reflection. Teaching in a campus garden brings to light new metaphors for learning that potentially transform teacher education. As teachers and teacher educators decide how to include 'the environment' and principles of 'sustainability' into their lessons, this book combines theory and practice to explore, step by step, how teaching in the natural world changes how people learn and how they teach. In this book, the lessons of building a campus Learning Garden with student teachers is described through three metaphors: garden as environment, garden as community and garden as transformation. Each metaphor is described in the contextual narrative of building the garden, as, gradually, teaching moves from an industrial, transmission model that emphasizes learning 'products' to a transformative, 'eco-centred' model that invites process and participation. The book also includes a chapter on 'practical matters' that provides starting points for developing curriculum in an ecological design model; there is also a list of activities, resources and links for further research. The book is by no means an exhaustive analysis of current theories in environmental education. Instead, this textbook is intended to introduce student teachers and teacher educators to current and historical ideas in environmental theory while providing a bridge between theory and practice.

As a starting point for eco-centred teacher education, this book thus attempts to describe how a garden brings a balance of teaching and learning that is at once practical and reflective, local and global and promotes teacher education with heart, and earth, in mind.

Walking the Talk

On the first day of the environmental education class, we talked about the role of learners in a society and economy dependent on diminishing natural resources. We talked about our place as teachers in an industrial model of learning that promotes production, completion, results and individualized notions of success. We talked about the possibilities of breaking a cycle of dependence on unsustainable models of life and learning. I told students that we would build a garden on campus. They were elated. We walked out to the site, which was covered in weeds, scrap metal, broken glass and other discarded junk. The students smiled. This was it. Freedom. A break from the four walls of learning. Under a clear blue sky, this piece of land would answer our questions. Inside the classroom, we had talked about issues that seemed insurmountable; out here, our thoughts were clear, problems seemed answerable. Even if we did not solve the larger problems of the planet, out here, we could learn about the environment, even as we created a new space, and process, for such learning. The Learning Garden began as a place to learn about environment, from within environment. It felt new.

I now imagine that if we had stayed inside and talked about building a garden, we would have imagined all of the difficulties in funding and other logistical challenges and might never have actually developed the space. In some ways the garden answered our questions even before we asked them; in fact, the garden presented us with new questions from the very start. Before we could learn about the environment, we decided to change how and where we would learn; before we decided what we would do, we needed to see what was already there. Orienting our learning around the new learning environment turned out to be a core principle for the garden's initial design. A space first conceived as a 'backdrop' for what we already knew turned out to be the centre of learning. At every step, the garden guided what we learned and how and where we learned it.

On the first day, the students looked at the barren patch of land, imagining what the place could be and what it could teach. It reminded me of the blank slate a teacher faces on the first day of school in September when questions revolve around lesson planning, curriculum, classroom management, but still, there is a large grey area, a set of unknown complexities that even the most experienced teachers are unable to plan for. The garden presented itself as a great expanse of similarly unknown pedagogical possibilities. It was, so excitingly, not a place for knowing, but a place for learning.

The shift from knowing to learning and from product to process became a central part of learning for the student teachers who would arrive in the next cohort. Early on, the garden permitted transformed awareness of place, a balanced starting point for students schooled in an industrial model of education. Such balance, and transformation, was badly needed.

> Among the most daunting challenges of our era is the task of bringing about the transformation of consciousness that will be required if we are to move away from a culture predicated on consumption and the values of the market toward one that strives to balance human activities with the requirements of the natural world. (Smith 1999, p.207)

Later, cohorts of student teachers had mixed views about learning how to become teachers in a Learning Garden. Some were in heaven. Some were relieved that they would be able to directly apply their interests in social science, food politics, climate change and other creative interests to teaching. Some felt empowered with a kind of teacher training where they would have a hand in deciding what, why, where and how they would teach. Others felt uncertain at the idea of engaging in a brew of ecology, curriculum, and hands-on, project-based, cooperative learning methods. Some asked: 'How can I learn...if I don't know?' Some were suspicious ('Is everyone doing this environment thing?'), while one student told me: 'I hate nature.'

For the most part, the student teachers, most of whom have recently completed a variety of undergraduate degrees, look to me to teach them how to teach in much the same ways they have been taught how to learn. Learning in a garden, then, takes up the challenge to shift learning patterns by inviting students to learn how to teach by learning deeply, by becoming aware of their own patterns of thought, specifically through the lessons and patterns of nature. Ecology-based teacher training is not an impossible concept for students to grasp, and yet, to bring deeper conceptions of learning in process, through the natural world, into a student teacher's understanding, involves challenging years of indoctrination in abstract, detached, formulas for academic success. Learning in a garden changes learning through an invitation to step back, to learn and, most importantly, to allow learning to take place, in place. It is an invitation to work and learn, not more competitively, but with eyes open to the life and land that surrounds learning. Learning in a campus garden teaches students through practice and the development of common sense philosophy.

Working with students in the garden showed me how the concepts of slowness, simplicity and gratitude can become core principles of teaching and learning that bring depth, wonder and heart to teacher education. I learned that teachers must learn. A Learning Garden is a setting that allows such a radical, common sense idea to take root. In working in the garden with students, in the challenges of developing community, we learned to recognize that in focussing on the 'problems' of the planet, we tend to ignore the problems within humans that have created the ecological problems. As Orr (2005c) reminds us in explaining the term 'Ecological Literacy',

> The disorder of ecosystems reflects a prior disorder of mind making it a central concern to those institutions that purport to improve minds. In other

words, the ecological crisis in every way a crisis of education…an ecologically literate person would have at least basic comprehension of ecology, human ecology, and the concepts of sustainability, as well as the wherewithal to solve problems. (p.xi)

If learning in a garden is to promote a shift in education through teacher training, from single subject, decontextualized, methods-based approaches to interdisciplinary, ecoliterate learning with a focus on process, then learning in a garden must unfold slowly, with care, sensitivity and deep awareness of one's surroundings. Such a way of learning mirrors nature itself, as Rachel Carson (1962) long ago described in *Silent Spring*:

> It took hundreds of millions of years to produce the life that now inhabits the earth—eons of time in which that developing and evolving and diversifying life reached a state of adjustment and balance with its surroundings. Given time—time not in years but in millennia—life adjusts, and a balance has been reached. For time is the essential ingredient; but in the modern world there is no time. (p.6)

If learning unfolds as life unfolds, in process and over time, do we not stand a better chance to learn slowly, alongside the rhythms of life, with a sense of wonder and awe? To allow that nature might be able to teach us something about life as well as something about how to learn and teach? As one student in the garden responded on the last day of class, 'The garden is a garden. But it is not just a garden, is it?'

Concrete Principles

A campus garden, simply because of the fact that it is outside, on a horizontal plane, under a limitless sky, provides the three-dimensional conditions whereby students begin to think in different directions. In a garden students think upwards, outwards and, ultimately, downwards to the earth. A garden encourages a close attention to detail, in the context of the universal. In a garden there is simply no room for single source, abstract ideas in larger-than-life versions of 'success'. Knowledge in a garden arrives from a sense of place, as ideas from textbooks are tried and tested.

The basic principles we initially set out for the garden (100% organic; focus on native plants; rotating stewardship among cohorts) required animation through action and/or discussion beyond the lecture and the textbook. The first 'lived' lesson in the garden, then, was to be mindful in the translation of ideas conceived out of context, inside classrooms, inside laboratories, before making those ideas 'real'. In the garden we learned the importance of reflective garden design developed in context, in determining a water source, in deciding what to plant, in the challenges of developing a shared space for learners and 'users' of land; in short, every action in the garden was a challenge to bring mindfulness to our action. We learned to ask ourselves: Does this principle include practice? Does this practice include principle? Does the manifestation of our ideas and principles honour the wildness of nature?

The garden promoted wholeness for student teachers in learning how to teach and in learning how to learn. In promoting interdisciplinary learning, social and team skills, PBL, community access and preventative classroom 'management', the garden truly supports so many progressive ideals of teacher education, in practice. A garden offers students a chance to learn within the context of nature in a way where the goal of learning is not so much to acquire knowledge by 'taming' it to meet our own needs, but in honouring nature's intelligence where 'wildness' is an integral part of that knowledge. The poet Stanley Kunitz (2005) says: "The danger is that you can so tame your garden that it becomes *a thing*. It becomes landscaping" (p.78). Work in gardens with students involves practical knowledge; it also involves a sense of wonder, immersed in that wildness.

The land and our involvement with it then taught another important 'teacher lesson': letting go. We learned that the land has its own mind. The students learned how garden stewardship, like classroom teaching, should promote ideals of sharing over ownership. As the poet says, nature is not *a thing* to be tamed. The land invited teachers to involve themselves with care, and then to very carefully let go of the object of care. The garden taught the students how teaching, like love, like nature, does not always spring from the rational mind, but from a deeper understanding of virtue that requires service, hard work and responsibility. Orr (2004) comments on the sense of selflessness required in our approach to the land and learning:

> Hope, real hope, comes from doing the things before us that need to be done in the spirit of thankfulness and celebration, without worrying about whether we will win or lose. (p.210)

It was an important lesson for the teachers as they made the link between environment, community, service and transformation. However, while the lessons of the garden were large, and profoundly affected how new teachers approached teaching and learning, the garden itself remained small, a site to return to.

Small Matters

How does a small garden that promotes community and a sense of wonder prompt such a shift in perspective? The garden itself is a paradox in that it represents a simple, small, local solution to large, complex, global environment problems. Learning in the garden is immersion in dichotomy and difference as a bridge towards understanding and community. The garden is a new context for teacher education that, itself, sparks questions about the ways that higher learning has remained severed from the natural world. Orr (2004) reminds us that developing a 'sense of wonder', which Rachel Carson discussed in the 1960s, and, especially, avoiding the destruction of a sense of wonder, is a starting point for environmental education:

> there is the danger that education will damage the sense of wonder—the sheer joy in the created world...it does this in various ways: by reducing learning to routines and memorization, by excess abstractions divorced from lived experience, by boring curriculum, by humiliation, by too many rules, by overstressing grades...and mostly by deadening the feelings from which wonder grows. (pp.23–24)

The movement outdoors to the garden was prompted by a sense of urgency, by the need to connect students to nature in meaningful ways, by a sense that eco-centred teacher education is long overdue. Edmund O'Sullivan (1999), writing on what he calls the 'historical decline' of the earth, urges educators to approach their work in light of a dramatically changing context; or, as he bluntly asks, "What time is it?" (p.13).

Exposing students to land stewardship in higher education is, I discovered, something still very new. The students held a variety of undergraduate degrees, from a variety of institutions, had achieved very high grade point averages, and yet were totally unused to learning right on their campuses. Several student teachers mentioned to me, as we walked to the garden for the first time in the term, that this was the first time they had gone outside for any of their classes.

Another challenge was to link ecology to teacher education, to promote the idea that the environment matters, or that 'the environment' even *exists*. On one hand, students in teacher education are starting individual professional careers, but learning to teach also involves social, cultural and, increasingly, ecological imperatives. Climate change brings a social and moral imperative to consider the environmental impacts of all professional training. Student teachers, like any student entering technical or professional programs, must understand how their practices, lessons, methods and even their own consciousness affect the environment at the local and global levels. The work of teachers is profoundly connected to the world; the world can no longer be ignored in a teacher's work. Ecologically responsive teacher training must include a foundation for learning that draws on deeper connections with the earth. To teach, and to train others to teach, as if the planet matters, is an obvious, ethical, ecologically just way of promoting sustainable learning. With increasing scientific evidence about the effects of global climate change, there is now a moral imperative to teach teachers as if the planet itself were not simply a unit or a course elective within a larger system of learning that is severed from the natural world.

There is hope. Teachers, especially student teachers, are in a position to profoundly impact the next generation of learners left with a legacy of a changing planet. As a response to the moral imperative to 'do something' about the environment, we built a garden. This book describes how responses to large issues facing the planet can be small.

'Y detras de cada escuela un taller agricola...
donde cada estudiante sembrase un arbol.'
[And behind every school is found a garden...
where every student plants a tree.]

Jose Martí, Cuba (1853–1895)
(Cited in Desmond, Grieshop & Subramaniam 2004, p.43)

CHAPTER TWO

❀

Learning in School Gardens: Historical and Theoretical Overview

Introduction

Outdoor learning has long connected youth to environment in ways that are both practical and profound. From the philosophies of outdoor learning in Rousseau to experiential education in John Dewey and Maria Montessori to post–World War II Victory Gardens to the proliferation in campus gardens across North America, gardens have played a central role in the larger history of teaching and learning. In this time of climate change, and call to environmental action, campus gardens provide a teaching and learning focus for environmental education philosophy, local tradition and hands-on agricultural practice. As educators and students build campus gardens, and develop their own curricula and goals around these unique spaces, it is useful for new teachers to gain a basic, historical and critical understanding of the deeper roots of school gardens.

This chapter thus provides an overview of the historical roots of school gardens and Garden-Based Learning (GBL), the role of gardens in the context of environmental education philosophy and the ways in which historical practice and theories related to environmental education can potentially influence current teaching and learning practice.

The Roots of Garden-Based Learning

Since schools, curriculum and programming reflect larger school communities, a history of learning in gardens is also a history of interactions between teachers, theorists, curricula and the changing nature of the planet itself. UN researchers (Desmond, Grieshop & Subramaniam 2004) currently define GBL as "an instructional strategy that utilizes a garden as a teaching tool. The pedagogy is based on experiential education, which is applied in the living laboratory of the garden" (p.20). The same researchers note how, in the history of GBL, several diverse goals, practices, philosophies and interests influence the level of support for school gardens. In other words, as with other issues connected to schooling (i.e., curriculum development, school district policy, state standards), the context of a garden (or the absence of a garden) very much mirrors the social, cultural and political interests surrounding it. Academic reform and vocational interests, in the context of socio-political ideals, have also determined the level of support. In Cuba, for example, the *Pioneros* programme promotes school gardens for both the production of food for the community and as a place to promote ideals of community-based work and self-sufficiency. The history of GBL in North America shows how support for school gardens, as part of core basic education, has been far less consistent, as the inclusion of GBL is directly dependent on academic reform movements.

On the other hand, school gardens have long provided an alternative to models of schooling that emphasized an industrial metaphor for school programme development in the early 20th century (Lincoln 1992). Reform movements in North America in the early 20th century, for example, were a response to schooling where schools were removed from society, in their own "distinct, walled communities" (Rothstein 1993, p.8). From a Western European perspective, GBL is rooted in philosophies of experiential education in the writings of Jean-Jacques Rousseau (1712–1778), Friedrich Froebel (1782–1852) and Johann Heinrich Pestalozzi (1746–1827), who believed that outdoor learning created balance between "hands, heart, and head" (Desmond, Grieshop & Subramaniam 2004, p.2). In the 20th century, Maria Montessori (1870–1952) and John Dewey (1859–1952) supported learning within nature as a means of both understanding and nurturing organic systems.

As the leader of the Progressive Education movement, Dewey believed students must move beyond textbooks and authoritarian teaching methods and engage in tangible practices that aid their overall well-being. In advocating learning as a process of developing common understanding, Dewey (1902) believed that education is the foundation for the larger society of learning: "A society is a number of people held together because they are working along common lines, in a common spirit, and with reference to common aims. The common needs and aims demand a growing change of thought and growing unity of sympathetic feeling" (p.14). Dewey added that, in an ideal education, learning would take place beyond the classroom, where "The life of the child would extend out of doors to the garden, surrounding fields, and forests...in which the larger world out of doors would open" (p.35). While Dewey did not describe the cultural or learning value of gardening *as work*, and while his goal was often based on an individualized student success model, from a Western approach to knowledge and schooling, Dewey's practical theories made important links between historical daily life, the land and learning. He said:

> Gardening need not be taught either for the sake of preparing future garden-ers or as an agreeable way of passing time. It affords an avenue of approach to the knowledge of the place farming and horticulture have had in the history of the human race. (cited in Subramaniam 2002, p.2)

School gardens flourished at the turn of the 20th century in Europe and Australia and were seen as a means of promoting the ideals of 'progressive education' and traditional notions of stewardship. Almost a century ago, in 1910, a report to an agricultural education conference estimated that there were approximately 80,000 school gardens in the United States alone (Welby 2003). At this time, school gardens were promoted in the United States mostly as an aesthetic feature in urban life. By 1918, every state in America and province in Canada promoted school gardens as part of growing food for the war effort; Victory Gardens thrived after World War II, after which school gardens declined and the land previously used for gardens was taken over by athletic fields (Subramaniam 2002). For some time in the 1960s and 1970s, school gardens rose in popularity again, in part due to liberal social move-ments. At the time, developmental psychologists also favoured experiential education and hands-on learning; in *The Ecology of Imagination in Childhood* (1977), Edith Cobb concluded that "creative expression is rooted in a child's relationship with the complexity...of the natural world" (cited in Evergreen Report 2000, p.3).

In the 1980s school gardens were not in favor in North America due to increasing socio-economic conservatism and a growing emphasis on localized performance standards. In the early 1990s school gardens once again made a resurgence, as the American Horticultural Society held an influential sympo-sium on youth-based gardening. Then, in 1995, former California state superintendent of schools Delaine Eastin began a large-scale, state-wide programme titled 'A Garden in Every School'. Between 1995 and 2002, school gardens in California increased by 3,000 and their numbers continue to grow.

> When I first called for a garden in every California school in 1995, I knew educators would respond because I had witnessed the transformation that occurs with students and teachers as they worked in their school gardens...[w]hat I did not know was how relevant gardens would become to the educational and nutritional challenges that have since emerged. (California Department of Education Report 2002)

Eastin stated that resurgent interest in school composting and farming was "widening the circle of learning from the school garden to local farms...[providing]...a context for students to understand...family farming and agriculture as well as the environment they will inherit" (California Department of Education Report 2002). Eastin's Garden in Every School initiative was based on the following principles:

• Gardens can create opportunities for children to discover fresh food, make healthier food choices and become better nourished.

• Gardens offer dynamic settings in which to integrate every discipline, including science and maths, language arts, history and social studies, and art.

• Young people can experience deeper understandings of natural systems and become better stewards of the earth.

• School garden projects nurture community spirit and provide numerous opportunities to build bridges among students, school staff, families, local businesses, and community-based organizations.

• Links with school gardens, school food service programmes, and local farms can ensure a fresh nutritious diet for children while teaching about sustainable food systems. (Cited in Desmond, Grieshop & Subramaniam 2004, p.26)

Alice Waters, chef and proprietor of Chez Panisse Restaurant, worked closely with Eastin and founded the Edible Schoolyard (TES), a one-acre, urban, organic garden at Martin Luther King Jr. Middle School in Berkeley, California. At TES, students plant and harvest food that is later prepared in the school cafeteria. The Center for Ecoliteracy (CEL), a public foundation founded in 1995, is closely associated with TES and acts as a large non-profit educational resource for teachers and resources.

Lessons in the CEL are focused around the following components:

- A knowledge component connecting ecological content and an understanding of the patterns and processes by which nature sustains life.

- A real world context with a hands-on experiential project that provides for the application of ecological knowledge.

- A social dimension in which students, teachers, parents, and others are given meaningful opportunities to work together as a community.

- A sense of place connected to the land. (Murphy 2003, p.4)

With pedagogical goals rooted in interaction with the natural world, the experience of TES combines practical knowledge of farming and the cyclical nature of "food systems, from seed-to-table" (p.2) as a global concept for learning. A learning model based on application of practical knowledge of the land also potentially changes relationships around the garden:

> The approach supported by the Center for Ecoliteracy and embodied in the Edible Schoolyard calls for a shift in relationship between teacher and student, adult and child, human and land, and is firmly planted in the context of a nurturing environment. (p.2)

The goal of the pioneer school garden at TES is to work efficiently and knowledgably with the earth's shrinking farmland and to shift focus in the total learning environment: from teachers to students to the land in a way that unifies the whole learning community.

Two of the founders of TES describe the goals of the garden this way:

> How do we cultivate in children the competencies of heart and mind that they will need to create sustainable communities? How can we design schools as 'apprenticeship communities' that model the practice of living sustainably? (Barlow & Stone 2005, p.1)

The TES model thus conceptualized learning as both a tangible outdoor experience and as a site where the land itself influences the shape of teaching and learning. In this way, TES directly challenges previous methods of teaching and learning through location, pedagogical goals and in the connections to larger, natural systems.

As other researchers at TES note,

> Small changes have profound effects. For instance, growing their own food in a school garden can open students to the delight of tasting fresh healthy food, which can create an opportunity to change school menus, which can create a system wide market for fresh food, which can help sustain local family farms. (Capra 2005b, p.27)

The socio-cultural impact of gardens in communities is also a vital element in the story of school gardens in ensuring that these sites survive for future generations. UN research notes the impact of gardens on non-English-speaking immigrant parents at the Evergreen Elementary School in West Sacramento, California: "This project raised the self-esteem of the children as well as the non-English-speaking parents who were then valued as teachers" (Desmond, Grieshop & Subramaniam 2004, p.42).

While a school garden can seem 'revolutionary', GBL easily adapts to mainstream schooling at a basic level. The rise in the school garden movement coincides with mainstream Project-Based Learning (PBL) methods, supported since the early 1990s, where teachers encourage self-directed learning and initiative. Research in PBL describes how students develop research questions and spend several weeks investigating interdisciplinary subject matter and synthesizing research (Curtis 2001). In a way that connects GBL to PBL it could be said that "[Students] are willing to experiment, collaborate, and immerse themselves in new ways of communicating, learning, and getting things done" (Consortium for School Networking 2006).

In other research that potentially connects GBL to mainstream learning, a preliminary study (Ryu and Brody 2006) of graduate level environmental education students involved engaging students in PBL activities on how daily consumption contributed to the students' Ecological Footprint (EF).[1] The sample empirical study involved having students measure their EF at the beginning of the course and at the end of the course, with the goal of measuring potential changes in student behaviour. The researchers found that actively engaging graduate students in environment-based PBL activities over a three-month period had a positive impact on student behaviour. The significance of the study in the context of school gardens and teacher education student research was that the researchers found that student behaviour was strongly influenced by two key factors: that the students were engaged in hands-on, practical activities related to sustainability and environment *and* that the activities took place in a specific context.

In many ways, interdisciplinary GBL provides an ideal focus and inspiration for project-based work. Along with an increasing number of curriculum materials linking government standards to GBL,[2] school gardens are increasingly supported as a curricular focus in the official channels of school district policy and practice and in current pedagogical theory. In comparative, historical terms, GBL is enjoying a renaissance nearly a hundred years after its initial popularity in North American schools.

Learning in school gardens is situated in a long tradition of experiential learning, progressive education and 'ecoliteracy', with support from educators, theorists and communities. Given the rapid increase in support for school gardens, there is a need for teacher education and postgraduate learning that will offer continued support in ensuring that future teachers have access to ecology-based learning on an ongoing basis. The tenacity of the school garden movement could also potentially influence a widespread, post-secondary learning movement where campus gardens provide a learning laboratory where students and educators develop progressive approaches to teaching and learning through hands-on involvement.

School Gardens and Student Achievement: Overview

The evidence is overwhelmingly positive in the ways school gardens contribute to students' academic achievement and to their well-being. The Learning through Landscapes Project was a landmark UK study that helped persuade educators and researchers to consider the academic 'worth' of school grounds. The study found that the impact of school grounds, both positive and negative, have an impact on student learning and attitudes about their total school experience: "school grounds...convey messages and meanings to children that influence their attitude and behaviour in a variety of ways... Children read these messages and meanings from a range of signifiers, which frame the cultural context of the environment" (cited in Evergreen Foundation Report 2000, p.2).

A U.S.-based, nation-wide survey also examined attitudes about Environment as an Integrated Context for learning (EIC) (Evergreen Foundation Report 2000). The study examined the experience of students, teachers and administrators in 40 elementary, middle and secondary schools that include the natural environment as a "curriculum element" (p.4). Further to surveys and interviews on school programmes, standardized test scores for students involved in EIC showed consistent improvement in reading, math, science and social studies. In comparative studies, EIC students consistently outper-formed non-EIC students in the same subjects. As researchers note in surveys of eco-centred learning programmes, "when students get involved in outdoor projects, their level of engagement and enthusiasm for learning goes up, not down, and that this situation, for teachers, translates into significantly reduced discipline and class management problems" (p.18).

In another study, over 4,000 school principals were interviewed on the use of school gardens in academic instruction. The study noted a strong need for curriculum materials, teacher training, funding and nutritional knowledge in creating successful gardens (Graham, Lane Beall, Lussier, Mclaughlin & Zidenberg-Cherr 2005). Other researchers report similar findings, noting that the success of experiential, garden-based, project-based learning relies on further research and on the development of a ecology-centred curriculum framework (Desmond, Grieshop & Subramaniam 2004)

A study commissioned by the TES (Murphy 2003) examined the Center's holistic pattern of education for sustainability which includes ecological knowledge, cooperative school culture, immersion in a hands-on project and learning based on a sense of place. The study also investigated students' attitudes towards health, nutrition and agricultural knowledge. Findings included a variety of positive outcomes, including the effects of school gardens on academic achievement and positive impact of the notion of *education for sustainability* on student psychosocial understandings. Some conclusions of the study indicate improvements in "understanding garden cycles", "psychosocial adjustment", "student sense of place", "understanding of sustainable agricul-ture" and "increased knowledge of the food students eat" (p.6). The study also noted that the teachers who taught in the schoolyard "ranked teaching academics, cooperation, and compassion for living things as their three highest priorities" (Murphy 2003).

Research commissioned by the CEL(Stone 2005) reported that

> ESY students showed greater gains in ecological understanding, and greater overall academic progress, than did students in a comparable non-ESY school. Teachers reported better behavior in class. Students who gained most in ecological literacy were also shown to be eating more fruits and vegetables. (p.229)

While more research is needed in the area of the impact of GBL on student learning at all levels, early studies show that learning through gardens improves overall academic achievement and student engagement. UN researchers (Desmond, Grieshop & Subramaniam 2004) strongly support a global need for more eco-centred curriculum, for further teacher training in garden-based methods, for more community involvement around environmental issues and for further study on the academic and social effects of GBL.

For student teachers and other educators new to GBL, it is helpful to know that proven, measurable results are achievable in school gardens. Providing evidence from such studies is also useful in helping administrators and educators gain programme support for school gardens at the district level. Still, for school gardens to gain widespread acceptance as a site for critical, educational change, educators (and their institutions) must also approach their gardens as more than a campus beautification project or outdoor learning site that acts as a curriculum tool for achieving standards. Even the measuring of 'success' in a garden perhaps needs to be rethought, as principal Comnes (2005) of Martin Luther King Middle School states:

> Can you prove that it's making a difference? I don't know that you can. What the Edible Schoolyard has done is helped change the culture of the school. It grabs kids' interest in school by giving them a hands-on experience with their peers and their teachers. It pulls kids into the school, into learning. And it makes kids realize that learning isn't just books, but that life is about learning. (Smith & Comnes 2005, p.147)

The difficulty in measuring the actual 'worth' of school gardens is that learning in a garden seems to somehow change evaluation criteria. My experience in the Learning Garden on campus, for example, taught me that teaching in a garden tends to emphasize the learning process over measurement of 'outcomes'. In school gardens, learning outcomes and standards may be outwardly achievable, but the real value of learning in a garden is even more difficult to measure and requires further research, further action and more narratives, in a variety of interdisciplinary forms. As I describe in the story of the Learning Garden, a garden brings forward entirely different definitions of success. Perhaps meaningful research in school gardens must adapt to measure more subtle 'success' in ways that will also redefine, even fortify, the notion of academic success. And so, while the above studies tend to support GBL in schools, measuring the 'measurables' only scratches the surface of the deeper, potentially transformative value of school gardens.

School Gardens in the Context of Environmental Philosophy

A school garden promotes learning as process in constant interchange between theory and practice. Viewing school gardens through a lens of environmental philosophy provides educators with a critical background that helps inform their decisions and discussions surrounding the garden. What writings in environmental education will help introduce undergraduate student teachers to environmental theory and inspire them to understand how their smaller actions impact larger systems? How does the (relatively small) work of a garden fit into the (larger) historical and environmental theory? How can new teachers explore eco-centred methods that contribute to learning in context, in a way that potentially deepens their understanding and reduces ecological impact?

Environmental education revolves around concepts of learning in context and relates to ideas about learning as interdisciplinary process. Environmental education approaches learning and knowledge in terms of potential ecological impact, instead of simply in terms of scientific knowledge advancement (Orr 2004). In such a model, environmental justice, community responsibility, land ethics, human ecology and measurement—such as EF—are all taken into consideration as part of learning. Ecology in education emphasizes land and learning, in context; as Canadian ecologist Stan Rowe (2006) describes, "Ecology is the science of context, a persistent reminder of the importance of what stands outside every object of interest. Ecology teaches the absolute dependence of things on what is peripheral to them" (p.34).

Even though 'the environment' has received more media attention of late, environmental education is deeply rooted in a traditional, international, theoretical context. For example, some German states encouraged school gardens as long ago as 1818 (Jewell 1908). In 1869, Austria and Sweden were the first two Western countries to officially initiate school garden plans into school policy, and both implemented plans in the same year. The Austrian imperial school law of 14 March 1869 prescribed that "where practicable a garden and a place for agricultural experiments shall be established at every rural school" (p.23).

The links between environment, education and schooling were first widely included in Western education institutions by Sir Patrick Geddes (1854–1933), who, in 1902 in Edinburgh, Scotland, promoted outdoor education with the idea of educating the whole person in the context of the surroundings (Palmer 1998). Geddes was also an early visionary in developing an interdisciplinary perspective for environmental education: "Nature study is not a new subject, to be squeezed into already crowded programmes, but the symptom, and in great part the leaven also, of the progressive transformation of these" (Jewell 1908, p.9). There was wide support in the academic community for Geddes' higher ideals around the 'revolutionary' integration of learning and environment. As Professor C.F. Hodge stated at the turn of the past century, "Nature study is learning those things in nature which are best worth knowing, to the end of doing those things that make life most worth living" (p.9).

John Dewey's (1902) theories on experiential learning also overlapped with environmental education, as Dewey believed that teaching and learning must always involve firsthand experience since "The environment is always that in which life is situated" (p.141). Early on, Dewey urged educators to embrace interdisciplinary learning because "analysis of isolated detail of form and structure neither appeals nor satisfies" (p.140) and that learning was most effective when experienced by "connection with life itself" (p.142).

Both Geddes' and John Dewey's progressive models of outdoor education and experiential learning predated the first UNESCO environmental education conference in Tblisi, Georgia (1977), where a firm mandate was laid out that linked environmental education to interdisciplinary learning: "As an educational approach... environmental education...can permeate a range of disciplines, both traditional and new, as well as form the mainspring of many integrated courses" (Palmer 1998, p.9). The report also noted that "environmental education is regarded as the embodiment of a philosophy which should be pervasive, rather than a 'subject'" (cited in Palmer, 1998, p.9).

The conference also laid out a number of recommendations about the future direction of both teaching and study in environmental education, including the ideas that environmental education

- is a lifelong process;

- is interdisciplinary and holistic in nature and application;

- is an approach to education as a whole, rather than a subject;

- concerns the inter-relationship and interconnectedness between human and natural systems. (p.9)

The final report of the UN conference in Tblisi set out a series of broad recommendations called 'Three Goals of environmental education':

1. To foster clear awareness of, and concern about, economic, social, political and ecological inter-dependence in urban and rural areas;

2. To provide every person with opportunities to acquire the knowledge, values, attitudes, commitment and skills needed to protect and improve the environment;

3. To create new patterns of behaviour of individuals, groups, and society as a whole, towards the environment. (p.11)

The recommendations and goals of the UN conference show how the 'tradition' of environmental education was always very much rooted in inclusive ideals that promoted both the earth and the earth community 'as a whole'. The inclusion of social justice, the emphasis on interdisciplinary learning that reflected the interconnectedness of both social and ecological systems, the inclusion of urban and rural areas and the recommendations for equality and holistic learning were different from conceptions of learning that promoted single subject, abstract, decontextualized knowledge. From the start, the goals of environmental education acknowledged the facts of ecological and social systems that have always surrounded learning. Such ideas foreshadowed Orr's notion that "All education is environmental education. By what is included or excluded, students are taught that they are part of or apart from the natural world" (p.12). In the past as well as the present, environmental education has long promoted open-ended perspectives that connect learning to 'the world'.

For environmental philosophers, interdisciplinary learning allows students to understand interconnections between academic subjects, while exposing them to the ways in which that knowledge impacts their surroundings on various levels. For example, in Orr's view, an interdisciplinary science curriculum should be linked to history, ethics, citizenship, globalization and, above, all, tangible, firsthand knowledge of how scientific knowledge impacts the world outside the classroom: "knowledge carries with it the responsibility to see that it is well used in the world" (p.13). He cites the *Exxon Valdez* oil spill, a tragedy made possible "because of knowledge created for which no one was ultimately responsible" (p.13).

Indigenous teaching and worldviews had a seminal influence on the evolution of environmental education thought throughout the 20th century. The first people of the Okanagan, originally known as Syilx, believed in approaching the land with an attitude of gratefulness, in respecting the land's capacity to produce food and water and enable survival, as a partner in the community (Marchand 1984). Within the framework of Traditional Ecological Knowledge and Wisdom (TEKW), providing young people with firsthand knowledge about native plants and their cultural importance is a central part of teaching, learning and living in harmony with the land (Turner, Ignace & Ignace 2000). In such a framework, the facts about the natural environment are learned as an inclusive system that combines ecological systems, adaptive strategies, respectful interaction and philosophy and recognition of (and respect for) the spirituality in nature (Turner, Ignace & Ignace 2000).[3] The appropriate use of TEKW, then, involves environmental study that cannot be separated from a deep, established ethic of respect and care for the whole earth community. In such a view, scientific facts, study or advancement cannot be included as valid knowledge without a land ethic.

Predating Western theories about experiential and environmental education, and current notions about sustainable resources and economies, indigenous education was the basis for environmental education learning principles that promote and nurture well-being in and for the earth. Such a perspective articulates systems of knowledge that include individual self-knowledge within larger context of family, the larger community and the globe. Okanagan educator, and director of the En'owkin Centre, Jeannette Armstrong says: "a community is the living process that interacts with the vast ancient body of intricately connected patterns operating in perfect unison, called the land" (Armstrong 1999). On the role of community as a cornerstone in developing respectful attitudes towards the environment, she adds: "Real democracy is not about power in numbers; it is about collaboration as an organizational system." And on the value of developing ecological awareness in community, Armstrong (2005) says:

I have noticed that when we include the perspective of the land and of human relationships in our decisions, people in the community change. Material things and all the worrying about matters such as money start to lose their power. When people realize that the community is there to sustain them, they have the most secure feeling in the world. The fear starts to leave, and they are imbued with hope. (pp.16–17)

Other current writings about environmental philosophy (Bowers 2006; Shiva 2005) discuss the reclamation of public space as a way of developing socially engaged, knowledgeable communities. Shiva talks about "reinventing citizens" and "living democracies" (p.84) that promote biodiversity, local action and provide a solution to a globalized monoculture and socio-economic injustice. Community-based projects such as school gardens also counteract the cultural impacts of larger industrialized societies and promote a shift in values, as Martusewicz (2006) explains, in the context of Detroit community gardens:

> In effect, the gardens are centers of community activity that foster a different set of values that are especially important to the younger members of the community: co-operation, learning how to nurture natural processes, acquiring the knowledge and skills that can be used to achieve greater self-sufficiency, developing good eating habits that contrast with the diet of industrially prepared food. (p.53)

A campus garden directly links to both traditional and Western philosophies of environmental education, by providing a space where students tangibly connect with the idea that local and global systems coexist within an organic cycle. Preservation of the rights of organic systems is a central part of ecological justice. Providing examples and methods for sharing spaces within a larger ecosystem begins in a garden with the sharing of local resources, which in turn protects the rights of the earth. For example, Shiva specifically identifies privatization of water as a symbolic threat to commons and community; instead of valuing nature's work and the organic, community decision making involved in sharing water and other resources, the privatization of natural resources turns an organic element into corporate property. Shiva (2005) points out an alternative: "No species seeks its entitlement to its share of water through the market place; they get their access to water through being members of communities and ecosystems" (p.43). In a garden students gain firsthand understanding of the ways that social and ecological justice and sustainability must be planned actively and progress alongside nature, not against it. Shiva explains how the integration of traditional knowledge is part of a new 'sophisticated' awareness that responds to a planet in crisis:

> Contemporary ecology movements represent a renewed attempt to establish that steadiness and stability are not stagnation, and that balance with nature's ecological processes is not scientific and technological backwardness,

but rather a sophistication toward which the world must strive if planet earth and her children are to survive. (p.51)

In the context of traditional and current environmental teaching, a Learning Garden is thus an attempt not just to teach students about nature or food or even politics but to help create a cultural shift that places the earth at the centre of life and learning, where "worldviews and lifestyles are restructured ecologically" (Shiva 2005, p.62).While a garden teaches students firsthand, traditional knowledge of farming and agricultural practice, it also provides a space where students tangibly connect with the idea that local actions connect to global consequences as linked ecological and social processes. A campus garden makes the links visible.

School Gardens and "Earth Democracy"

An important element of "Earth Democracy" is the idea that "living economies are built on local economies" (Shiva 2005, p.10). Shiva emphasizes balancing the needs and resources of local economic systems and notes that conservation and sustainable livelihoods all emerge from the local level, which in turn support national and global economies: "Localization of economies is a social and ecological imperative. Only goods and services that cannot be produced locally–using local resources and local knowledge–should be produced non-locally and traded long distance" (p.10).

When students at all levels experience the small gesture of building a campus garden, they are thus involved in a much larger awareness of how their food and energy choices connect and how their collective choices profoundly impact both local and world economies. It is, then, vital for students to be aware of both *what* they are learning and *how* they are learning in gardens. As Shiva says, "*Earth Democracy* connects the particular to the universal, the diverse to the common and the local to the global" (p.1)

Learning in school gardens specifically links to concepts of social and ecological justice in the way that a community-built and maintained space acts as a living example of a shared resource. In the promotion of biodiversity, organic growing practices, shared decision making and other eco-centred ideals, gardens help create the social, biological and underlying cultural conditions necessary for sustainability to take place.

Ecological security is our most basic security; ecological identities are our most fundamental identity. We *are* the food we eat, the water we drink, the air we breathe. And reclaiming democratic control over our food and water and our ecological survival is the necessary project for our freedom. (Shiva 2005, p.5)

Small-scale growing practices promote biological diversity in the cultivation of organic crops and in the free sharing of seeds. However, biological diversity is a scientific term that carries with it social implications that can help students develop a larger understanding of sustainability. In other words, it is not enough to promote scientific knowledge of the facts of sustainability; we must promote the underlying *learning* conditions and values that will make sustainability possible.

A Learning Garden is an example of Earth Democracy 'in action', as it promotes community, moderation and humility in a grass-roots, alternative learning model. In a garden, sharing and working the land as a community, combined with local and global awareness, promotes both ecological justice and greater understanding of human rights as they exist within a local and global, sustainable system.

As technology increasingly shapes knowledge, as land development increasingly encloses opportunities for learning in community, a community-built Learning Garden is an alternative for learning that moves outwards, past the enclosures, supported by the grass roots. As an Earth Democracy the garden challenges globalization through work that promotes local, ordinary action and shared decision making based on respect for land in the larger context of the earth.

Earth in Mind

In his text *Earth in Mind: On Education, Environment and the Human Prospect* (2004), David Orr calls for the integration of environmental education across the curriculum and a science curriculum linked to history, environmental ethics, citizenship, globalization and firsthand awareness of how scientific knowledge affects the world outside the classroom. Such a curriculum supports the belief that "knowledge carries with it the responsibility to see that it is well used in the world" (p.13). Orr calls for a new kind of education across disciplines, an 'ecological design intelligence' that involves "the capacity to understand the ecological context in which humans live, to recognize limits, and to get the scale of things right" (p.2). Orr cites certain scientific and technical experiments described as 'breakthroughs' during the 20th century (i.e., war technologies, pesticides) that were invented in isolation, without concern for environment or human life. Instead of perpetuating decontextualized, abstract academic subject matter, learned and taught in isolation, Orr suggests that education itself should instead focus on developing relationships that unite subject matter, humans and the earth. He emphasizes the consideration of moral responsibility in how knowledge is used and acquired in the first place. Orr emphasizes approaching the land "carefully, lovingly, competently" (p.105) in ways that incorporate firsthand, practical knowledge that will be used well in the world.

Orr also questions entrenched, subject-centred, abstract models of academic learning and stresses the value of teaching through the "concrete reality" of "natural objects". In *Earth in Mind*, he discusses industrial models of education which produced non-sustainable economies, capitalist myths and visions of human independence, all of which contributed to an anthropocentric approach to learning and living that separates humans from nature. He criticizes historical systems of learning which severed the human need to interact with environment and, instead, he promotes 'ecological intelligence' and "the value of being involved in a context/environment you choose not to exploit" (p.44). His approach to environment and learning is rooted in context, in the interconnectedness of systems, and in the value of responding to problems with the planet while, above all, keeping 'earth in mind'.

In criticizing the nature of detached, scientific exploration with a sole emphasis on isolated factual information, Orr emphasizes that some responses to environment defy logic, and must do so:

> Science without passion and love can give us no reason to appreciate the sunset, nor can it give us any purely objective reason to value life. These must come from deeper sources. (p.32)

Orr's approach to environmental education (and his inclusion of values such as 'love' in relation to 'environment') is not so much based on sentimental visions but is very much linked to the nature of intelligence itself. For Orr, learning deeply within the context of environment presents a very real, human, organic possibility of working respectfully within nature, where students learn to separate the trivial from the important and to think clearly about the world beyond school. He notes how the standard of intelligence now is based on "how well we market to jobs, advancement, economy" whereas "true intelligence is long range and aims toward wholeness" (p.11). He cites Thomas Merton, who described the purpose of mainstream education as the "mass production of people literally unfit for anything except to take part in an elaborate and completely artificial charade" (cited in Orr 2004, p.11). In seeking harmony between humans and the earth, he calls on educators to consider the value of instinctual, emotional attachments for the earth and cites Stephen J. Gould's famous phrase: "We cannot win this battle to save species and environments without forging an emotional bond between ourselves and nature as well—for we will not fight to save what we do not love" (p.43).

In provoking attention towards the illusion of a 'status quo' model of learning, where the goal of scientific 'advancement' and 'increased knowledge' is an unquestioned cornerstone for learning, Orr advocates that humans respect all that we *do not* know and allow ample room for doubt while making decisions that will impact the planet:

> The imperative is...we must pay full and close attention to the ecological conditions and prerequisites by which we live. That we seldom know how human actions affect ecosystems or the biosphere gives us every reason to act with informed precaution. (p.xiii)

Orr calls for 'ecologically literate' practices in teaching and learning and a return to the idea of *education* from the Latin root, where *educe* meant 'to draw forth'. As firsthand experience of the land defines, Orr, in a tradition of environmental education that goes back to the first UN conference in Tblisi, calls for 'interdisciplinary' education and "breaking free of old pedagogical assumptions, of the straight jacket of discipline-centric curriculum, and even of confinement in classrooms and school buildings" (p.33). Orr ultimately emphasizes learning in (and *as*) the inseparable context of the earth when he states, "All education is environmental education" (p.12).

The Garden "Commons" and Ecological Justice

Besides the idea that a school garden illustrates Earth Democracy and embodies an ethic of caring necessary to the earth's survival, other concepts related to current and traditional ideas about the environment are useful in helping to broaden and define research and practice in school gardens.

A garden provides a space where students tangibly connect with the land in a way that permits a practical, profound understanding of the local and the global. As I learned time and again in the Learning Garden, a garden is a garden; at the same time it carries metaphors for complex awareness that might be difficult to translate from theory to practice. Current ideas in environmental thought promote the idea that we need to move beyond the idea of 'the earth' as a 'unit' of study and encourage a much more profound respect for nature. New writing around environmental philosophy more and more reflects traditional ideas (and indigenous worldviews) and promotes a larger concept of ecological awareness where the earth is a living entity, with equal rights, and where the preservation of those rights is a central part of social and ecological justice.

In a garden, protecting the rights of the earth begins in the most tangible way, with discussion around local resources. Shiva (2005) identifies the need for discussion around shared resources with the symbol of water privatization as a threat to commons and community; instead of valuing nature's work and grass-roots decision making, Shiva says that the enclosure of water and other resources automatically turns an organic element into corporate property. In a garden, students begin to understand how social and ecological justice and sustainability are planned actively and progressively, keeping in mind the needs of nature. In this way, while a school garden can be called a 'progressive' model of learning, there are profound implications in the way gardens promote 'traditional' ways of life. Shiva describes:

> Contemporary ecology movements represent a renewed attempt to establish that steadiness and stability are not stagnation, and that balance with nature's ecological processes is not scientific and technological backwardness, but rather a sophistication toward which the world must strive if planet earth and her children are to survive. (p.51)

School gardens specifically link to concepts of social justice in the way that a community-built and maintained space acts as a common resource and provides a living example of how to share resources. Promoting biodiversity, organic growing practices, shared decision making and other eco-centred ideals, gardens help create the social, biological and underlying cultural conditions necessary for authentic sustainability to take place. In this way, a garden becomes Earth Democracy that "connects people in circles of care, cooperation, and compassion instead of dividing them through competition, and conflict" (Shiva 2005, p.11).

As a form of 'the commons', a school garden provides students with an example that "embodies social relations based on interdependence and cooperation. There are clear rules and principles; there are systems of decision-making" that also include "a democratic form of governance" (p.21). A school garden, though small, makes visible non-dominant patterns of social organization.

Writing extensively on the subject of 'the commons',[4] Bowers (2006) argues that Western philosophy, based on a long academic tradition of rational thought, has severed human connections to the land, the commons and our own sense of survival "as the lives of people in the West become even less centred on the self-sufficient possibilities of the commons, and more in the industrial culture that is beyond their control, their insecurity becomes more palpable" (p.8). Bowers suggests that the 'hubris' created by traditions of philosophy that deny the natural world has dangerously led away from traditions and mentorship that would promote survival and integrity of humans and earth:

> The result is that [students] will not be aware that the democratic process should involve the interplay between critical thinking and a deep and complex understanding of the traditions that are the traditions of the basis of civil society, the community patterns of mutual support and self-reliance, and the conservation of the commons. (p.21)

Bowers argues that 'progress' or 'change for change sake' is no longer a viable goal when it comes to ecological justice or ethical encounters with the land. For Bowers, a university's clear role and responsibility is to 'revitalize the commons',[5] since

> The threats to what remains of the world's diverse cultural and environmental commons represent a unique challenge to Western universities, particularly since what these universities have designated as high-status knowledge has played such a dominant role in undermining both the cultural and natural commons. (Bowers 2006, p.vii)

Like Orr and Shiva, Bowers talks about the disregard for tradition and traditional ways of life rooted in environmentally irresponsible academic practice. Revitalizing the commons, permitting students to take part in traditional land practice and learning in/as community, paradoxically, becomes a 'new and progressive' revolutionary act.

Revitalizing the commons, then, is a means of stemming the tide of isolated individual or economic interests that seek to expand, harness or altogether destroy natural elements in a way that harms larger community:

Enclosure, and what has come to be called 'the tragedy of the commons' where a member of the community attempted to expand his herd, fish catch, or other practice, at a rate that, if everybody were to follow the practice, would overwhelm the sustaining capacity of the natural systems. (Bowers 2006, p.3)

The concept of the commons and revitalizing the commons takes place in amazing, unexpected ways in the context of public school and community gardens. Martusewicz' (2006) work alongside inner city youth in Detroit community gardens shows how many facets of 'the commons' (land, language, culture, tradition) are open to revitalization. She states how the Detroit commons are "a testament to the fact that even the poorest of the urban poor can envision an alternative future—and can mobilize the community in a way that leads to less reliance on the industrial culture that is failing so many groups" (p.48).

Learning in school gardens, students are not simply following a teacher directive or mastering a curriculum goal, but are breaking out by defining and redefining what learning in a local, shared place really means, based on their own unique experiences.

In the garden, even knowledge is shared, not owned, as the garden itself becomes, even grows into, the embodiment of that shared knowledge. The methods—the teaching and learning that takes place in school gardens—are rooted in community and traditional knowledge. Because community and tradition have largely been ignored in teaching and learning during the latter half of the 20th century, a farm model of learning, is, paradoxically, seen as 'progressive' and 'alternative'. As Martusewicz says, gardens "viewed within the larger context of the ecologically and cultural destructive impact of the industrial culture...represent expressions of resistance" (p.48).

Future Directions: Limits and Possibilities

In the teachings of Orr, Shiva, Bowers and others, ecologically just learning is always rooted first in the natural world. Current research stresses the need for both practical and critical understandings in teaching in school gardens and the need to further examine concepts such as direct food, globalization and anthropocentric learning models (Edmundson 2006). Such a need can be realized through education programming that supports interdisciplinary, ecology-centred concepts with firsthand experience of land and food. The garden provides a place where students can consider, up close, the threats to local food sources through local real estate development; global agri-business; the commoditization of a basic life source (land and seeds) and various forms of embedded industrial knowledge and learning (mentioned above) that contribute to ecological damage.

UN researchers strongly support the development of more school gardens from both a practical and a pedagogical perspective:

> It is reasonable to expect that our current ideals of educating children through an integrated curriculum, dealing with issues relevant today, and recognizing the unique potential of every child could be practically realized through the stable establishment of school gardens. (Desmond, Grieshop & Subramaniam 2004, p.8)

If, as these and many other researchers suggest, learning in school gardens has such a significantly positive influence on students, why is GBL not widespread, and mainstream, in all North American schools? In addition to the idea that school gardens in North America tend to be directly linked to progressive reform, UN researchers have other potential explanations:

> One is that the pedagogy has not been critically examined and endorsed by educational researchers and practitioners...Related to that shortcoming is the lack of infrastructure support for school gardens or related GBL efforts. Finally, there is often no local strategy to sustain the physical plant of the garden site as a permanent part of the school or programme facility. While school athletic facilities often receive significant school and community investment there are few examples of similar support in the fields of environmental education or GBL. (p.8)

Researchers also note that school gardens are often underestimated in their significance in the development of student learning that perhaps school gardens are predominantly viewed as little more than a teaching tool that transmits curriculum in standard methods of instruction.

In the story of our campus garden, described at length in this book, the garden was sometimes viewed by academic colleagues as little more than a landscaping enhancement project. Some dismissed the garden as a 'nice idea' that held relatively little impact on the direction of teaching and learning in the field of education. Others saw it as proof that 'sustainability' existed on campus. And, some understood the deeper learning benefits, how the garden directly links students to learning and community and to the natural world in ways that dramatically shift how students previously encountered the learning process itself. Such a diverse range of perspectives *all* shaped the growing narrative of the garden and also tended to influence how we approached the garden, in the larger context of the university.

Given the variety of perspectives on learning in school gardens on schools and campuses, researchers have lately begun to examine whether ecology-based teaching and learning involves both a radical change in method and perspective towards the learning process.

> Are green school grounds initiatives instances of ecological transformation integrated with other efforts in transforming pedagogy and wider society, i.e. pedagogical and social transformation? Or, are they, perhaps (still) something 'extra' and 'outside' of the mainstream and therefore, while important for ecological transformation...[contribute] little to changing the social and pedagogical status quo in schools? (Dyment & Reid 2005, p.287)

There are now calls for more study on the potentially deeper role of greening school grounds since "there has been little research exploring how green school grounds relate to the values and goals of the wider educational system, or if and how these processes contribute to a 'sustainable education'" (p.287). In a study of two high-profile school greening projects in England and Canada, researchers explored how gardens potentially create social transformation, described as "a change in personal behaviour and social and organizational practices" (p.287). The study showed that the potential for "broader scale transformation to occur remains largely unrealized when there is a lack of institutional and structural frameworks and vision to support and nest these initiatives" (p.287). Researchers recommend that larger-scale change is necessary in teaching and learning "for green school grounds projects to become truly transformative" (p.299); they suggest that such projects "forge closer links to the socio-political and environmental learning agendas of citizenship and sustainability education" (p.297).

Interestingly, so much of the latest evidence around environment and education tends to conclude that in order for students to develop closer connections with the natural world, change on a much deeper level of consciousness is required. While there are differing notions of how to attain this 'deeper level', researchers agree that some form of attitude shift is required:

> Moving away from an educational system largely driven by economic concerns to one aimed at preparing children to shape and sustain an ecologically

beneficent society will require a constituency of adults willing to support a 'green' curriculum and 'green values'. This suggests that if ecological education is to become widespread, adults and children must learn how to interact differently with both the natural world and its human inhabitants. (Smith 1999, pp.208–209)

As school gardens take root, and researchers, educators and others consider different ways to promote an 'alteration' in consciousness, can colleges of education and teacher training in particular nurture new narratives for eco-based learning? What is the role of teacher education in helping to create links to the broader, societal goals for sustainability?

There are some examples on North American college campuses [6] where educators have developed ecological awareness in a way that includes the entire campus community, inside the classroom and out. However, while college campuses increasingly incorporate the term sustainability in their goals and mission statements, there is work to be done in incorporating values around ecology and community into schools of education and teacher training. In fact, teacher education programmes are potential campus leaders in connecting concepts of environment, teaching, learning and sustainability. Bowers describes the progressive role for colleges of education that include cross-campus collaborative efforts in ecology, community and culture that "provide students in-depth exposure to ecologically and community oriented faculty both within education and other departments on campus" (Bowers 1999, p.173).

UN goals for global sustainability (UNESCO 2005a) place education in a central role. The United Nations Decade of Education for Sustainable Development (DESD [2005–2014]), for which UNESCO is the lead agency, describes the need for a combination of both practice and theoretical innovation in the work of global sustainability. The goals of the UN DESD include integration of "the principles, values, and practices of sustainable development into all aspects of education and learning" and education that "encourage changes in behaviour that will create a more sustainable future in terms of environmental integrity, economic viability, and a just society for present and future generations" (p.6).

The goals of the UN DESD should be of particular interest to teacher education. One of the more ambitious goals of the UN DESD, for example, is to bypass the typical '10-year lag time' between educational innovation, research and practice. In fact recent recommendations prepared for the UN

DESD make special mention of the role of teacher education in furthering ESD practice:

> One major group that is worthy of special mention in terms of capacity-building and training are teacher educators along with pre-service and in-service teachers. Through many contact hours in the classroom, the world's 60 million teachers mould the knowledge base and worldviews of millions of children. If pre-service and in-service teachers learn to weave ESD issues into the curriculum and to use pedagogical techniques associated with quality ESD, then the next generation will be capable of shaping a more sustainable world. (p.19)

The goals of the UN DESD are as far-reaching and link to the larger goals of environmental education and school garden programmes:

> Education for sustainable development is about learning to: respect, value and preserve the achievements of the past; appreciate the wonders and the peoples of the Earth; live in a world where all people have sufficient food for a healthy and productive life; assess, care for and restore the state of our planet; create and enjoy a better, safer, more just world; be caring citizens who exercise their rights and responsibilities locally, nationally and globally. (UNESCO 2005a; Vision & Definition of ESD)

The goals of DESD as set out by the UN specifically emphasize 'change' to existing education programmes and 'training' to achieve such change. The four main goals of DESD are focused on "access to quality basic education; reorienting existing education programmes; developing public understanding and awareness; and providing training" (UNESCO 2005). The DESD also includes goals and methods with a strong focus on "processes of public participation for integrating indigenous, traditional, and local knowledge and culture into ESD programmes" that "help learning become more transformative in nature" (p.20).

Specific methods for promoting education around the theme of sustainable development mirror larger, historical goals of school gardens. The UN mission statement for DESD encourages teaching and learning methods that involve the following characteristics:

•   Interdisciplinary and holistic: learning for sustainable development embedded in the whole curriculum, not as a separate subject;

- Values-driven: sharing the values and principles underpinning sustainable development;

- Critical thinking and problem solving: leading to confidence in addressing the dilemmas and challenges of sustainable development;

- Multi-method: word, art, drama, debate, experience...different pedagogies which model the processes;

- Participatory decision-making: learners participate in decisions on how they are to learn;

- Locally relevant: addressing local as well as global issues, and using the language(s) which learners most commonly use. (UNESCO 2005)

For many environmental researchers, if teachers and students are to be involved in the health and future of the planet, there is an urgent need for a paradigm shift in teacher education that begins with the earth. The UN indicates a global mandate for education based on a universal vision of democracy and equal rights for the earth and developing countries; meanwhile, school districts struggle to meet the challenge of incorporating sustainability into their teaching programmes. Teacher education can play a part in bridging theories around global ideals and local, tangible practice. Teachers, especially in their early training, can bring the priority of the survival of the planet into the classroom through learning that promotes ecological awareness combined with experiential, hands-on practice.

More and more research (Louv 2008; UNESCO 2005a; Corcoran 1999) supports the idea that engaging teachers in the environment, especially in teacher education, is a potentially powerful way to shape student learning around ecology. The 'urgent need' of 'reshaping' teacher education will also bring about the most dramatic global impact in 'reshaping' attitudes towards environment:

> One of the challenges of the coming decade will be reshaping teacher education to acknowledge the far-reaching changes that must be made if our culture is to become ecologically sustainable...If we are to fulfill the need for education for ecologically sustainable action, we must prepare undergraduate pre-service teachers through environmental education courses. (Corcoran 1999, p.179)

Other current research cultivates ecological awareness in student teachers through autobiographical writing activities that honour a student's historical connections to their surroundings:

> By reacquainting future teachers with the powerful impact that encounters with the natural world have had in developing their own sense of themselves, they become more likely to seek ways to provide their own students with experiences that will instil a love for that world in the next generation. (Corcoran 1999, p.179)

Even small changes in student programming can make a dramatic difference:

> the setting of even a single course, if thoughtfully and carefully constructed, can provide students with the opportunity to examine the purpose and meaning their education and their capacity to put it in the service of their environmental concerns. (p.187)

Much of the current research around ecology and education describes how deeper changes often begin in small, subtle ways. Changes in thinking towards environment do not just begin with lesson plans, policies or standards. One of the founders of TES explains the unquantifiable link between values and practice in developing 'ecoliterate' attitudes:

> At the Center for Ecoliteracy we believe that at their heart, the ecological problems we face are problems of values. We've noticed over the years that it's very hard to change the values of adults, while at the same time we've noticed that children are born with certain values intact—namely their sense of wonder and their affinity for nature. David W. Orr reminds us that the biologist E.O. Wilson calls this 'biophilia'. We all share that trait, but it seems particularly strong in children. It's undiminished when they're young. And one of our philosophies is that we think that, properly nurtured, biophilia can develop into ecological literacy and eventually lead toward a more sustainable society. (Barlow & Stone 2005, p.45)

How can we train teachers to nurture the hearts of their students? Evidence shows that as the earth warms, as the demand increases for the integration of environmental education in all subject areas, as the number of school gardens continues to grow, teacher education must prepare future teachers in methods that include a global environmental mandate and local, contextual awareness for the natural world. Such 'environmental immersion' invites students to acquire hands-on eco-centred methods, to read current environmental philosophy, to translate such ideas into practice and, in short, to transform the teaching and learning process when teachers first learn to teach.

In order to nurture the ideals of sustainability and 'the environment' in new generations, teacher educators can encourage new teachers to be emboldened by their gardens, which are always, at once, practical and visionary, local and global. One way to nurture the heart in our practice is to help turn our students towards their natural surroundings and life. Eco-centred teacher education is a place to begin.

Towards Learning in/as Place

> As a child, my escape to another world was a swamp near our house in London, Ontario…Every day in that marsh I could always count on finding something new; some exciting new creature or world to discover. Today, that swamp is entombed by a huge parking lot and shopping mall. The vast diversity of life has been replaced by an enormous array of consumer products. What does that mean for youths who spend their time there now? (Suzuki 2002)

The uniqueness of a campus garden in the body of environmental research is that learning in a garden involves a large number of students in promoting sustainable practice (i.e., farming local land) while also developing deeper awareness of concepts like Education for Sustainable Development, 'the commons', Earth Democracy or other ideas related to ecological justice. Simply put, a garden has the power to translate complex theory into practice. A garden in teacher education thus helps deepen pedagogical practice by reaching new teachers who will bring their ideals around environment to their schools.

Historically, theoretically, locally, globally and in everyday learning practice, school gardens have evolved as a site that helps students become more knowledgeable about the natural world. Whether viewed as 'progressive practice' or as a 'return' to traditional values around the land, a school garden prompts a cultural shift where the earth is central to life and learning, where "worldviews and lifestyles are restructured ecologically" (Shiva 2005, p.62). While a garden teaches students agricultural practice, a garden provides a space where students tangibly connect with the idea that the global and the local profoundly coexist. A garden helps students recognize both the visible and the invisible connection, and interdependence, between humans and the earth. The uniqueness of a Learning Garden in higher academic settings is that conceptual ideas become visible, making it that much more difficult to ignore the fact that natural species depend on the earth, and each other, in contexts both large and small, for mutual survival.

A survey of theory and practice in school gardens and environmental education supports the idea that garden learning potentially opens students to a wider world of understanding, one that can help address local and global environmental concerns in practical ways. In light of changes to the earth's climate, pedagogical research overwhelmingly points to the need for teaching and learning in the context of environmental education in a way that moves beyond a few 'green lessons' or a single 'ecology unit' in a science methods course, and more towards transforming environmental consciousness. Based on both scientific evidence and moral obligation, caring for the environment must become a new focus of the largely unspoken agreement of responsibility and trust between the public, schools and teacher education. Current research also brings new questions to light: With the current climate crisis, what will be the next generation's contribution to the planet? How can educators inspire students to relate to the natural world as the natural world changes? How can educators bring the deeper recommendations of environmental theory (i.e., interdisciplinary learning/community mindedness/holistic methods/global concerns/garden-based learning) to mainstream learning?

While environmental researchers now urge greater critical depth in garden-based teaching and learning, the examples of educators, over the course of centuries, developing practices and ideas rooted in learning that supports the natural environment, bring inspiration and hope. Given the resurgent popularity of and support for school gardens, given the current interest in ecology and sustainability at all levels of industry, perhaps a garden is a bridge that will make the ideals and visions around sustainability 'real'.

The following chapters in this book describe the ways in which learning in a garden develops ecological awareness, community, and ultimately, a transformative view of both teaching and learning that touches on all levels of education.

CHAPTER THREE

✿

Garden as Environment

Introduction

> There are no larger fields than these, no worthier games than may here be played. Grow wild according to thy nature...let the thunder rumble...take shelter under the cloud...Enjoy the land, but own it not. (Thoreau 1854, p.141)

How does environmental immersion in teacher education promote ecological ideals while transforming the teaching and learning process itself? How can a campus garden engage students and teachers in environmental philosophy while promoting new metaphors for eco-centred practice?

The Learning Garden was first intended to expose students to a concept of the land as both a physical space and an experiential learning process, concepts involving responsible land management, ecological justice, risk taking, community commitment and, ultimately, the transformation of land and learning. Through physical labour, collaborative learning and written reflection, the garden inspired students to become teachers with heart, and earth, in mind.

The Learning Garden began with a land stewardship committee, a small group of graduate students (working teachers) and undergraduates (student teachers) in an environmental education course. The first group made a number of decisions for the garden, including the idea of a rotating steward-ship model and a focus on drought-tolerant plants. The garden then passed through a series of cohorts of student teachers and graduate students. The environmental education class would meet in the summer to plan, to develop policies and vision, to build infrastructure and to attend to watering. The student teachers met in the fall term to harvest and, after a 13-week winter practicum, they planted the garden during a four-week research project, the capstone of their programme.

The practical process of immersing students in basic, organic growing practices—the garden and garden stewardship—started an unplanned, profound transformation that seemed to touch all of the cohorts involved in the garden's evolution. By engaging with their community, and the land, students began to question their deeper attitudes towards nature and learning. Some students were able to apply practical or scientific knowledge to the garden, including the composition of soil or progressive irrigation methods; most of the students were able to take part in the shared application of that knowledge. They experienced how knowledge is also transformed through dynamic, organic, community processes. They learned to learn in response to Wendell Berry's approach to ecological design intelligence: "What is here? What will nature permit us to do here? What will nature help us to do here?" (cited in Orr 2004, p.105).

A community learning model, with garden work at the core, promoted local and global knowledge of drought, food systems and farming practices. A model of working and learning outdoors inspired students to want to acquire and share such knowledge in the first place. The garden shifted learner awareness from personal achievement to the environment itself; from student stewardship of the garden to the impact of that stewardship beyond the garden and into the world.

Among brand new student teachers, students just beginning to think about their own philosophy and teaching style, learning in the garden challenged assumptions of student/teacher 'success' and also some of the ideals of environmental education. It was especially the challenges that helped align their ideals and exposed them to the unpredictable processes of both teaching and the natural world. Learning in the garden, both as physical space and as a conceptual model, challenged the roots of teacher education.

In this chapter, the critical challenges of building, teaching and learning in a campus garden are described through the metaphor of garden as (physical) environment. The garden as environment, a literal outdoor space, involved practical knowledge of local climate conditions and the necessity for drought-tolerant plants and native species. An awareness of the garden as environment also promoted concepts of ecological and social justice, with, for example, the decision to donate produce from the garden to the local food bank. As time passed, learning in the garden constantly shifted between metaphors of environment and community, as students began to experience how transformative teaching and learning lies within awareness of the learning process itself. The students came to this subtle lesson mostly through a process of trial and error and often through setting the goals for their own learning. However, the 'new' idea of 'learning to learn' also arrived through traditional techniques, especially through cooperative mentorship, and by listening to elders' knowledge and experience. We recognized that the garden was built on traditional knowledge, on basic ideas about how to plant and help things grow organically and well. This setting provided a place for future students to practice traditional techniques and, in the process, to reflect on the ways that nature and community shape how we teach and learn, in the past, for the present, and as a gift for the future.

From the start, the garden was built on a practice of inclusion and equality: the inclusion of nature's lessons, of traditional growing practices and nature itself. Such lessons were a dramatic contrast to traditional teacher training where students are 'taught' how to plan lessons in isolation from their students and surroundings. The combination of shared, practical, inclusive knowledge, in written and oral reflections, became a central, profound routine in the garden. The students found their voices as teachers in a way that included, and respected, their surroundings.

While we learned that eco-centred learning might begin with a lesson plan around GBL, or PBL activities, deeper changes in behaviour and attitudes took off beyond the lesson plan, in unplanned ways. Truly experiential teacher training had begun. Instead of standing in a sealed classroom trying to convince pre-service teachers that 'learning takes place beyond the lesson plan', the students were already there, in the garden, experiencing it for themselves.

'Transformation' was never an explicit goal of learning in the garden, but was a subtle gift that arrived through immersion, through giving to the land, and in being sensitive to our surroundings. We discovered how transformation in a garden begins with concrete practice, rooted in a specific place and ultimately moves to the recognition of a learning process that is sometimes difficult to name because it is rooted back in practical experience itself. Just when the students thought they had moved into a transformative phase of their learning, they were back to weeding the garden, back to the practical demands of responsible land cultivation, the 'beginner phase' that would start the learning process again. In this way, the students and I learned how a garden also transforms teacher education; instead of control, a garden brings spontaneity; instead of fixed knowledge, the garden promotes a cycle of new understanding. However, these transformations took place later. First, there was the garden.

Garden as Environment: Why, What and Where

> Water, soil and the earth's green mantle of plants make up the world that supports the animal life of the earth. (Carson 1962, p.63)

> Those things nearest at hand are often the most difficult to see. (Orr 2005b, p.88)

The garden began as a site that would provide a context for 'hands-on' learning about urgent local environmental issues. The area surrounding the campus garden, the city of Kelowna in the Okanagan Valley of British Columbia, was once a small resort town in a fertile agricultural region surrounded by farms and apple orchards. The area is now a rapidly growing tourist, golf and wine region where the rising price of real estate pressures local farmers, many of whom have lived in the area for generations, to parcel or sell their land. Between 1976 and 2003, for example, the human population of Kelowna, an agricultural centre of approximately 88 square miles, increased by 94% (compared to a 35% growth rate in Canada during the same period) (Pidwirny 2002).

In addition to population increase, an important local ecological issue is the pressure on water systems in the arid desert climate. Much of the irrigation system originally set up by fruit growers is now used for golf courses, wineries or for real estate development. In the surrounding valley, 70-80% of water is used for irrigation (Cohen et al. 2006). The local valley has very little annual rainfall and the average warmest temperatures in all of Canada; the area is prone to radically reduced water supplies and intense, prolonged drought. The potential impact of water shortages on the golf and wine industries is significant and will eventually impact hydroelectricity production and salmon habitat, two core economies in the province. Another environmental issue in the area is wildfires; in summer 2003 wildfires destroyed 250 square kilometres (61,776 acres) of vegetation on the surrounding mountains. With warmer winters the Japanese pine beetles continue to destroy, en masse, the area's characteristic pine forest.

In such a context, there is an obvious need, in fact, a responsibility, to involve students, especially student teachers, in understanding the conse- quences of human impact on the environment and the interconnectedness of natural systems. Current research notes that human survival in fact depends on such knowledge. A 2005 UN report on the impact of climate change says that planetary and human survival in the midst of a changing climate depends on 'adaptive capabilities',[1] which include developing knowledge, attitudes and willingness to engage in environmental issues among local populations (United Nations Environment Programme [UNEP] Annual Report 2005). In light of both the surrounding context of the campus garden and current research, teacher training seemed like the natural place to develop adaptive capabilities through the garden.

Despite all evidence, and the obligation to make 'ecology' a central part of teacher training, the idea of promoting change in teacher education seemed like a daunting prospect. Teaching has, historically, managed to largely maintain separation from the natural world. Curriculum is typically delivered as a single subject in an abstract context. Ecology-centred teacher education, an interdisciplinary approach with 'ecology across the curriculum' as a key strand across all levels of programming, is virtually non-existent in North America. Why?

Naturalist and educator Paul Krapfel (1999) suggests that the separation of learner from environment in teacher education is supported by a lack of context-specific materials for training teachers about their local environment. He also says that there is simply a general lack of interest among many teachers for bringing students into the daily experience of the local, natural world: "Although the activities are about nature, they could be done in such a way that students remained completely isolated from the natural world. In working with teachers, I also learned that most lacked knowledge of the world around their school" (p.48). He adds that "most [teachers] had never learned to view the natural world with intellectual wonder" (p.48). Krapfel's last comment touches on a complex pattern of academic conditioning that has influenced teaching and learning for centuries. Why, then, have teachers 'never learned' to include their actual surroundings in their teaching? How has the notion of 'intellectual wonder' so long remained associated with abstract discoveries, described through textbooks, taught in isolation? The reasons for academic bias towards single subject, abstract, context-free learning are discussed at length by David Orr, who explains how the academic tradition itself promoted a separation between intellectuals and their environments since the time of Francis Bacon.[2]

The challenges run deep for starting ecology-based teacher education in a campus garden. One such challenge is the socio-economic context of the campus: the surrounding area is a semi-rural, geographically isolated, fossil fuel driven economy and culture that is heavily dependent on imported goods. Transportation for the areas is largely centred on one main highway that runs through the interior of the province. The need to provide a model for responsible land use within eco-centred teacher education was obvious; it was also obvious that one of the main barriers in developing adaptive behaviour in this and other 'car and mall' regions in North America was the historical lack of interest in planning for small-scale, sustainable, grass-roots projects.

Programme development that supported ecology and teaching towards environment would, in the early stages, be a somewhat lonely task. Since the campus garden began, national media coverage on environmental issues and the science of climate change has increased; meanwhile, a dedicated group of community gardeners in the area has helped community gardens flourish and school gardens to begin. However, the question remained: how could the teachings of a small-scale organic garden hold pedagogical or professional relevance, or promote adaptive capabilities, within an industrial culture focused squarely on fossil fuel? As we began the garden, in all its smallness, such capabilities seemed almost impossible.

Building a garden in such a context brings forward more questions about the scale and goals of learning that teachers have become accustomed to. What is the role and responsibility of the university in role modeling values and policies that support the scale of nature? What is the university's role in developing visible, perhaps smaller, examples of authentic care for the land that veer from the intellectual tradition of separation of learner from environment? Given the enormity of the goals around developing a sustainable planet, are small-scale solutions even credible? Without the university as a gathering place for asking questions and progressive thought, where will suburban and rural learners learn about concepts like local and globalized food production, about their EF, or even, about the impact of drought beyond their own front lawn?

While greening campus grounds is often held up as a participatory model of sustainability, there remain few narratives as to how campus gardens and school greening actually contribute to sustainability or to thinking sustainably

(Dyment & Reid 2005). If the UN document (2005) suggests that hope lies in participatory, grass-roots models of action, and, as noted earlier in this book, if many researchers now call for a deeper shift in attitude towards environment and learning, what are the actual adaptive capabilities presented by a campus garden? What is the connection between teaching, learning and the land? How does a garden supply a new narrative of the changes that might take place in both land and learning?

In practical and philosophical terms, the valley needed a campus garden, a place within higher learning that would provide a site for students to gain firsthand exposure to urgent local issues like the effects of drought, the relevance of native plants and the need for xeriscape as an alternative to perfect green lawns. The leap was taken. The idea of *garden as environment* took root. Senior administration supported the idea and, a small, out of the way parcel of land was designated as a garden cultivation site. That summer, the students and I simply showed up, without any funding, with one (borrowed) shovel, and with high hopes to 'do something' about the environment. When we arrived at the site, there was a small hilltop of rock hard clay covered in weeds, abandoned junk and debris, and the prospect of a garden seemed like a dream.

We started weeding. Conversations began. Our initial decisions were guided by our work; the hardiness of the weed roots in the clay soil helped us decide to build raised beds. Students with knowledge of soil began talking about ratios of sand, soil, manure and compost. Other students began investigating the geographical placement of the garden on a small hilltop: was it also man-made? Were the weeds native? When is a weed a native species? Some students sat in chairs to the side of the site, reading texts on environmental theory, engaging others in discussion. Everyone arrived at their own role. Did we have to 'do' anything with this land? Was the decision to cultivate this land even responsible, even if we aimed for responsible land management? Was leaving the land 'empty' a better decision for the environment? Even if no one had claimed the land, we did not want to subscribe to the colonial concept of *terra nullius*,[3] or to the idea that empty land is 'no man's land'.

During the first garden discussion, around emptiness, terra nullius and the surrounding grassland, it seemed possible that, despite all, we could nurture this land as a space to be shared and honoured. In fact, our very first Learning Garden principle came to light: we wanted to 'leave space' in the garden to honour that space. It seemed right. We still had much practical and pedagogical knowledge to learn from the garden, so we also knew it was a principle that we could not so much immediately apply as grow into. Several cohorts later, an area we first cleared for the cultivation of flowers would end up being developed by a future class as grassland.

It was not until we began the experience of building the garden that we decided: yes, we would build a garden. Others had been there (junk and debris were heaped around the site) and we would cultivate the land as a model of sustainable growing practice for a drought prone agricultural region, rapidly filling with land development.

We decided that the Learning Garden would be an inspiring alternative in land use for the region, and in the way we made decisions, we would become a community of *learning to learn*. These and other insights always arrived in the doing. From the very start, it was in the making of the garden that the garden became what it was.

While at first the garden seemed impossible, an isolated slice of sustainability sitting by the side of the highway, a change in values slowly took root. In fact, the recognition of values as the foundation of developing ecoliterate attitudes soon became a key part of learning in the garden. We realized that we could begin with practical policies for the garden, but as the founders of TES (Stone, Barlow & Capra 2005) and others learned, a garden also needs to cultivate the heart, and a deliberate turning towards care. Such values are needed for adaptive capabilities or ecoliteracy to grow. Over time, student teachers began to speak, without embarrassment or shyness, about their *love* for nature, their recognition of the mystery and magic found in the garden, the profound change in themselves, their hopes for future students and even about their awareness that the garden might not always continue.

Meanwhile, we weeded. We continued to debate and discuss. Work in the garden in the first days was based on student skills and knowledge, on applied skills of farming, gardening and water use. Even after a few days, the entire classroom site was already very unlike any class I had ever taught in. Somehow, I imagined that one day we would return to the indoor classroom; we would sit at tables, watch PowerPoint presentations and raise hands in formalized discussion about 'the environment'. But, after one class, the students decided that they wanted to meet in the garden every day. The room booked for our class sat empty during summer, as the new Learning Garden took on a magic of its own.

For this and later cohorts in the garden, a simple routine evolved. Students met in the garden and engaged in writing activities in their Field Notes journals. I asked them to make connections between the hands-on work in the garden and how their minds and hearts translated that knowledge. Before we began weeding, the activity on the first day was *describe your very first memory of nature*. The simple sharing of stories, outdoors, gave me an idea of how different learning in a garden could be. One student wrote in the journal:

> In the shady spot at the south end of the plot, we sit, reflect, and share some personal significant nature experiences...I remember thousands of outside adventures.

One student decided to write about the pond beside the garden; his first remembered experience as a child in nature was, to our everyone's amazement, Walden Pond. In the garden, it was easy to see the connections between early experience, current ideals and the spaces that people were drawn to.

After telling stories, we continued shovelling.

> I did not read books the first summer; I hoed beans. (Thoreau 1854, p.79)

Garden as Environment: The Challenge

In the first cohort, a student teacher/farmer/beekeeper in the group had strong ties to land that his family had cultivated consistently for over a hundred years. With rising land prices, he was pressured to sell off more of his farmland to economic development. The student immediately took to the concept of building a garden as it mirrored his own values. He wrote in his journal:

> As a farmer and an educator, my dream is to be able to bring agriculture/gardening into the classroom...or more to the point to bring the classroom to agriculture/gardening...When I found out that this class would revolve around the concept of a learning garden, that we would actually have the chance to design and work on one...I was thrilled.

The day we arrived at the garden and saw the hard piece of land covered in knapweed and filled with piles of metal, glass and other debris, one student commented:

> Admittedly, my vision of the garden site [before seeing it] was, perhaps, a little hopeful...Upon further reflection, however, I think it is quite fitting

because any site that one might get in a schoolyard would not be a plum spot, but rather one similar to what we have here...a spot that no one else wants to use and is out of the way.

Another student, a teacher/botanist, was also a former forestry expert who provided the group with knowledge of local soil, plants and climate:

Our first look at the garden site. It has gravely Ponderosa Pine, bunch grass... The soil is straight moraine...there is knapweed, mullen and spear grass everywhere. Also blue bunch wheatgrass, arrow-leaved balsamroot, cheat grass, yarrow, star-flowered false Solomon's seal, Tortula ruralis, Cladonia, a couple of Douglas-fir on the north pond exposure, and lots of ponderosa pine.

While we weeded and talked, we decided on some guiding principles and conceptual approaches for the garden: sustainability (working the land for future learners); interdisciplinary learning; hands-on learning (learning by doing); xeriscape as alternative to green lawns (responding to local water issues); organic (a contextual awareness of our surroundings as ecological systems); local traditions (community-minded teaching and learning); rotating stewardship (respect for future, changing needs of the land). The means of developing the garden's principles were also meant to create a tradition of discussion that would be passed on to future groups, who could discuss, change, solve or adapt the founding principles. All decisions were made openly, in the context of our new Learning Garden community. In order to cultivate the land sensibly, from a variety of perspectives, the basic plan was for a food/drought-tolerant/flower mix that would create a blend of 'beauty' and 'use' while showing how native, non-native and invasive species responded to drought.

The 'learning outcomes' and 'evaluation' were tangible. If the flowers and vegetables withered due to a water shortage and the xeriscape plants lived, students would have a visual example of the effects of drought. The plan also included scrapping it if it no longer made sense given (as yet unknown) future ecological conditions in the garden and its surroundings. In short, it was always important to give the garden back to the land. The plan was never to create a showcase for local plant life, or to promote individual student success, but to support a community learning process where mistakes brought understanding. We also wanted the actual garden to take place, organically. Even though we were not yet sure what the specific lessons of the garden would be, we also wanted to learn the universal lessons of the garden; we wanted to find the 'learning' in the Learning Garden, which we soon realized was centred around both practical skills and community. In a non-competitive, community model of learning, the lessons would last, even if, we realized, the physical garden did not. A non-ownership model, a shared space for learning and principles to match these core values set up the conditions by which eco-centred eco-minded learning might be possible. Even by recognizing the value of place in learning, by staying in tune with the needs and nuances of our surroundings, we were already changing how we learned. It would be a new, valuable, sometimes difficult lesson for new teachers.

Meanwhile, in the local business community, the idea of a Learning Garden took hold, and our funding challenges were solved almost effortlessly. Local businesses stepped up and made donations. The first donation was from a local lumber yard, which donated one thousand dollars of red cedar for raised garden beds, with promises of supplying more at wholesale prices. A local sign maker from the local farmer's market offered to carve a sign for the garden, which later became a painting project for a student teacher in art. Although local employees of national hardware chains wanted to support the garden, we were consistently turned down by the head offices of these franchises, which were located at some distance in Eastern Canada and the United States. Instead, the owner of the smallest nursery in the town began freely giving us plants. (She told us that since her children were now adult professionals, a lawyer and a doctor, she wanted to 'give back' her 'investment' to the campus.) Other local businesses in the small community immediately recognized the value invested in a campus garden; they loved the idea that an academic campus could provide a place for many community members, including their own children, to learn. They recognized how even the idea of a campus garden could motivate and inspire students, and help train new teachers in a way that they also valued. The Learning Garden caught the community's imagination and had an immediate effect of teaching people about the connections between sustainability and 'education' through direct involvement.

In building the garden infrastructure through shared community support, we learned the value of not waiting for official, single source funding before starting a project. In waiting for funding approval, there is always the risk that official channels might not yet include grass-roots, community-based knowledge or behaviour change as core definitions of sustainability. In the early days we also learned, first-hand, how a university's tangible mission and goals can be carried out, in fact, enhanced, by educators and students in hands-on, community-based projects. I found that working with new and experienced teachers in the garden that first summer built ideals of ecological justice and community momentum around the environment. We learned how working towards sustainability must include values (respect, justice, community, sacrifice, trust and care) that are also sustainable.

Some time later, another local nursery donated 300 flower bulbs; some bulbs were planted in the garden boxes while the rest were taken by a small group of students to the forest. The following spring, flowers started shooting up all over campus. Clearly, a new setting for learning was already taking shape.

> You begin to reshape your days to respect a more natural world view. You learn how to merge with the cycles, seasons and snows. Your hands get dirty and your back gets sore and you love it, because you know it means you're truly alive. Count on all this to not only sharpen the significance of your everyday routines but—stay with me here—to put you on the right side of history in the most important battle of our time: the struggle to determine how we'll all live together in the cities of the future. (Tracey 2007, p.21)

While the flowers planted by the 'guerrilla gardeners' took root in the hard ground that winter, we secured a grant from a major Canadian bank.

The following year, we received another.

Local Justice

After a class field trip to the local farmer's market with lunch that included a salad with cherries pollinated from the bees of our student farmer/beekeeper, the students considered the value of maintaining local farms as a way of challenging global food trade, especially in the context of so much imported produce in local grocery stores. During the conversation we pondered different scenarios. Like many local farmers, the cherry grower was elderly; what would happen if she sold her land to developers? What were the land ethics, the issues of eco-justice involved in building large-scale, permanent condo developments on fertile agricultural land? What was the connection between a local garden and globalized food ethics? Could we raise awareness about globalized food by building a campus garden in teacher education? How could students involve themselves in knowledge of global food by learning and working in school gardens?

Our work seemed very small in the context of global food politics. However, as Shiva describes in *Earth Democracy*, recognizing the flow between local and global connections is part of a larger responsibility. By building an actual garden on a post-secondary campus, we were beginning to create alternative representations of land use in the region as well as develop new representations of academic work that included hands-on response to ethical responsibility. The local area was full of examples of large condominium developments with advertising slogans written in bold letters on giant billboards. One billboard close to downtown read: "The World Revolves Around You"; another billboard on a large-scale development on drought-prone lakefront farmland read: "Water Without End". One look around their surroundings told students how agricultural land in the region was bought, sold and developed for non-agricultural interests. The students could see for themselves how a great deal of energy and advertising went into blurring the connection between land and food.

> If you put the earth's water…into a gallon jug, just over a tablespoonful would be available for human use. Ninety-seven percent of the planet's water is in the ocean, and two percent is locked in ice-caps and glaciers. (Hass 2005, p.108)

What is the use of a little garden? What can a garden, as a model for local land use, teach about globalization and responsible land practices? We pondered the questions from practical and philosophical angles, as gopher and deer attempted to build new routes through the construction that surrounded us.

On the blackboard, we forged some basic connections:

- Land = food;

- The issues of land and food are micro (the slugs are eating the tomatoes) and macro (how to get rid of them?) and interconnected (how do our decisions affect our local/global surroundings?);

- Small gardens help students understand the energy and effort required to farm and feed the world;

- School gardens teach students that an alternative exists to imported, outsourced, unsustainable produce sold at large grocery franchises;

• If students take part in the production of food, from seeds to watering to harvest, they will experience, first-hand, the life and possibilities in their local land;

• Encouraging students to plant and harvest inspires a sense of wonder in learners, teachers and communities;

• Teachers can promote critical awareness of local and global food sourcing by having students consider concepts that promote traditional ad local knowledge and small market economies;

• The combination of hands-on work in the garden and exposure to critical concepts helps students make a connection between local, organic, sustainable farming as both a valuable applied skill and as new knowledge;

• Land=life.

A totally organic approach to gardening, without pesticide, or even inorganic compost, led to larger understandings of our work in the world.

One of the student teachers commented:

How cool it would be to not only expose students to the atmosphere, but as farmers themselves...fruits and veggies that they could grow in their own school garden at a market like this! Showing students an alternative to the farmer's market, for example, with a trip to a big food emporium that provides food at high cost (to the consumer) and lowest cost (to the marketer). Then we could have students ask, "Where do these (New Zealand) apples come from?" That would help to define the globalized food process. We could ask...what are the costs for global trade?

The local garden provided a practical context for their readings in globalization from previous course work and personal interest. Some student teachers arrived at the garden nodding their heads in understanding, as if the garden embodied much of their previous reading in social and political science. The experience of speaking with local farmers, combined with planting and growing food in their own garden, provided a new sense of connectedness between academic understanding, daily practice, social values, love of the land and, now, teaching.

The campus garden flourished into a mutually supportive community. Beginning with the simple concept of 'a garden', the student teachers devised ways to bring this knowledge to their own classrooms. Ultimately, discussions linking local and global concepts through the shared space of the garden challenged forms of teacher education based solely on performance standards, organizational abilities and classroom management. Teacher training in a Learning Garden also taught important teacher qualities of humility, empathy and recognition of the responsibility of the privileged status of professionals in the developed world. In such a context, expectations for professional and personal success changed; the expectation of things not 'turning out' as planned, in allowing something even better to emerge in a lesson, was also a lesson for the new teachers.

Our work was an example of care, vulnerability and commitment to values, within the walls of academia. If, as Orr says, "being vulnerable requires a degree of agility and openness that is alien to the enclosed curriculum of the academy" (Orr 2005c, p.103), then the garden uniquely helped student teachers acquire teacher qualities of agility and openness from the very start of their teacher training.

The idea that we could start locally, with a small garden, while expanding to include global concerns, connected students to the larger world, made them want to be 'good citizens'. While the garden nurtured a new kind of teacher training, the students also challenged previous ways they had been trained to learn.

One student described the process of 'learning to learn' in the garden and how the new recognition of scale and place not only brought new lessons, but new ways of learning old lessons:

It is definitely also about...letting go, moving on. The Learning Garden creates better learners, better living. It [the course] seemed to evolve into values education, but, in a good way. You could never teach values education or any of the other stuff the same way again after the Learning Garden...local/global food issues... the garden does make you think about it all.

Practical Wisdom

One student in particular, the farmer, played a large part in teaching the other students about the value of process in teaching and learning. On one of the garden work days, two local youth set fire to the farmer's hay bales, which forced him to sell off his entire cattle stock. The student teachers noticed the student farmer's cool and effective approach to loss: he took each problem in the context of a much larger problem to solve. In fact, facing each problem, firsthand, in context, without worry or abstraction, seemed to give problems a chance to *be* solved. The students admired the student farmer's method of working: he rose early and worked long days to take part in the garden community; he set both his mind and body on a task and followed through; he listened to community members, allowing them to influence the intensity or passivity of his work; most importantly, he made decisions, then accepted and worked through the consequences of those decisions. He could easily admit his decisions and actions were wrong, through the process of working the land.

The teachers, uniquely, learned how to become teachers alongside a farmer, at work. The farmer's example of patience, flexibility, humility and hope exemplified certain teacher traits that are often difficult to teach in teacher training. The farmer learned and taught without colonizing the minds of student teachers or replicating the values of the teacher educator's own academic and teacher training.

> What does it mean to be educated and by what standard is that mysterious process to be judged? (Orr 2005a, p.104)

The question strikes at the heart of the notion that authentic teaching and learning can exist within organic process. Student teachers in later cohorts would write about the way the garden taught them to teach through the processes of life and equitable values:

> I wrote my last field notes about what I have learned from the Learning Garden The lessons were simple: you can't control weeds any more than you can control other people or everything that happens to you. Even with the best planning, gardens, like life, have a habit of going their own way...much of our wisdom comes from the natural world. We learn about growing and nurturing and even dying from the world in which we live. Nature is an equal opportunity teacher if we respect her.

Student teachers in a garden learn from a basic premise that learning is not so much a solvable problem, but an interactive process. Learning lies in learning that process. In the narrative of teaching teachers in the garden, a question presents itself: what is the teacher's role in the process of learning? In the garden we realized that what a teacher brings of value to a classroom is the recognition and facilitation of a learning process within the context of actual, organic surroundings. Such facilitation is brought about by the very qualities that drew all of the other students to the farmer in our midst: namely, the values of wisdom, humility and adaptation.

> For children, most importantly, being in the garden is something magical. As one of our teachers put it, "one of the most exciting things about the garden is that we are creating a magical childhood place for children who would not have such a place otherwise, who would not be in touch with the earth and the things that grow." You can teach all you want, but being out there...that's an ecology that touches their heart and will remain important to them throughout their lives. (Capra 2005a, p.xv)

Learning to teach alongside a farmer brought knowledge, wisdom and a unique awareness of the limitations of human effort that comes from the ebb and flow, the cycles of working the same land, year after year. From the farmer, students also learned about hopefulness as a practicality, that the notion of building a place for future generations to grow and learn is far from sentimental. In fact, I frequently noted how hope fuels the physical and social demands placed on people working closely together over days and weeks. The farmer provided a firsthand example of how the environment teaches the learning process. Many students began imitating the farmer's simple, effective habit of attending to the process itself; of working to the end of the day, making decisions in the context of the environment and others and returning the next day to begin the process again. When energies lagged, the farmer inspired everyone to return the next morning. As the process of the Learning Garden itself began, the metaphor of the garden as environment began to include the garden in the context of the larger environment, in an adaptive community of learners.

I remember one moment in particular, mid-way in building the garden the first summer. We still had a lot of hard physical labour ahead of us on the garden infrastructure. Learning had turned from discussions around environmental philosophy in the previous days towards more of a daily physical grind. Exhaustion was setting in. We had been shovelling sand all day in the hot sun, and at one moment we heard the call of the red-winged blackbirds in the pond next to the garden. We all stopped working and listened.

There was a rustle in the nearby pine woods. It was a hot day and everyone was tired. The farmer stood up straight and said: "Hope keeps the farmer going." We agreed. Then we kept shovelling.

I thought we shovelled a bit more vigorously after that.

The Pond: A Great Stirring of the Depths

> Education, I fear, is learning to see one thing by going blind to another. One thing most of us have gone blind to is the quality of marshes. (Leopold 1949, p.158)

> God made the sun...Then after that, he could see. All water. Nothing but water. (Syilx Elder Harry Robinson, cited in Blackstock 2001, p.3)

The garden is located next to a pond filled with a variety of migratory ducks, red-winged blackbirds and other wildlife. One of the first ideas that someone suggested when we arrived at the site was to use the pond to water the garden. The idea was to rig up a kind of pump that would transfer the pond water to the garden boxes.

That evening, one student wrote in the nature journal:

> I'm worried about the water issue. Disrupting the pond to water the garden? How will this negatively impact the real anomaly of a riparian habitat in this desert?

The student's concern for the pond raised more questions for the other students. What was the environmental impact of draining the pond? How did we interfere with goals for long-term, sustainable land and water use by removing water from the pond? Why was our first impulse in moving towards sustainable land management to destroy it? What previous learning had led us to seek short-term gains, while destroying other life forms? Leaving the pond alone seemed like an obvious, ecoliterate choice; however, the process of coming to this decision was the first instance where a practical need led directly to questions of environmental ethics.

The shift from seeking solutions to asking questions about ecological justice began with contextual awareness, occurring organically within community, within the decision-making process itself. The issue of the water made us realize that eco-centred decisions require a constant, conscious effort to weigh the ecological impact of human actions within an ethical framework of ecological justice. We realized that you could not always follow first ideas, as they were often based on previous, destructive, industrial or colonial patterns of learning. The students learned to think about a problem, then rethink the problem, several times over; learning with earth in mind did not come naturally. Water held importance for its own sake, and not recognizing that sacred importance was leading to a shortage of water in our region. The campus could talk about sustainability, and we could talk about our role in working around the water in the pond, but when it came to action, where was our common sense?

> Water is a meditative medium, a purifier, a source of power, and most importantly it has a spirit. Water is alive—biotic. (Blackstock 2001, p.5)

The article about First Nations perspectives on water in the Okanagan talks about researchers' visit to the Nlaka'pamux elder Millie Michell, who discussed the ways that water is important to traditional peoples. The article also discusses how, for the first several minutes during the interview, Michell wanted to know why they were interviewing her, since she believed that water was an important issue for all people. She kept asking, "Why do they...come here to ask us about water, isn't water important to them as well?" (p.5).

From our first experience with water in the garden, we learned that while water was needed to help the garden grow, it was not valued as life itself in the context of land stewardship. An industrial model of learning had taught us to sever our beliefs from our actions: on one hand, we knew that water usage in a drought-prone region required limits; on the other hand, we were ready to drain the entire ecosystem in order to accommodate our idea to cultivate the garden. If the garden died, would we simply call the whole project an 'experiment' and move on? Our learning was not situated. How could students learn to teach in ways where their actions did not connect to their surroundings? How had this happened? Considering an academic tradition of decontextualized learning, of constantly wanting to consume without considering the impact of that consumption, the ecological crisis now seemed like it had always been inevitable. The garden, thus, was not simply a site for learning about the land, but for learning about our interaction with the land and the impact of previous learning on the land. The garden taught us what we needed to know, before we began learning.

A traditional Hopi perspective describes how "water can be given and it can be withdrawn by the 'very something' that creates and sustains all life" (cited in Blackstock 2001, p.4). In such a perspective, and worldview, the Hopi say that the underlying causes of the water shortage is "a consequence of improper spirituality, which shows up in irreverent—that is, ignorant or greedy—interaction with the earth mother" (p.4). Local elder Mary Louie states: "The water, it's a gift of life. It bothers me because our water is...disappearing because it's not being respected" (p.4). Within the context of an indigenous worldview, leaving the pond alone was not a 'non-action' but a reverent gesture that valued both the biotic and sacred qualities of water. To drain the pond would have been to act on our own limited knowledge of these qualities. Our actions were guided by true respect.

It seemed right that our first symbolic action would be to listen to one another and to 'leave the pond alone'. The 'pond story' was the first we listened to, and we allowed that knowledge to change our actions. The pond was also the beginning of a storytelling tradition we still nurture in the garden.

> Water, we call it Mother Earth's blood, her nourishment to her children. I call this term 'the blood of life'...and without it we'd never survive. So we need water, and we need to keep it clean because if it continues in the manner that it's going...a person would wear a new pair of shoes right down to nothing before they'd get to clean water. That's one of those things that the ancestors talked about...we need to learn to preserve water. (Mary Louie, First Nations elder, cited in Blackstock 2001, p.4)

> You can't live without water, without the water we can't survive. And I can remember our Elders talking about it. Therefore, when we're weighted down with a lot of grief, your life is becoming unmanageable, or you're going through a lot of pain, the first thing our grandmother and my aunt and my mother would say, 'go to the water'. Water is powerful and yet it can be so gentle. You can see that when there's a big washout, the water can bring down boulders and big huge trees. It can move anything—a whole mountainside. And yet if you sit by a little brook, which I often did when I had a home up at Mabel Lake, I can feel that—I experience all my Elders taught me—I personally experience it. (Mary Thomas, First Nations elder, cited in Blackstock 2001, p.4)

The students learned that water is life and that learning in a garden brings an ever-widening concept of the natural world. The decision to leave the pond alone made the pond the sister of the garden in a metaphor of peace and care. The pond was a 'watershed' moment for learning about the complexity of natural systems, human learning communities and the interaction between both. There would be other stories about the pond in future cohorts. From that moment onwards, the pond refocused student efforts and helped to shape the community's values. From the garden, to the pond, the students looked further outwards, noting new metaphors in the massive campus development, the machinery and construction sites that surrounded our (suddenly shrinking) island.

> Water is local and global and complicated because different distributions of water and weather in different geographies have made for different animal and plant species, and human beings symbolize themselves through the plants and animals they live among—in one landscape raven is trickster, in another coyote. (Hass 2005, p.109)

By the following year, in the context of campus construction, the garden and the pond could be seen as either symbols of sustainability, or symbols of shrinking natural forest. In terms of teacher training, the students began to learn how to include metaphors for nature that flowed in their ideas; the spontaneous inclusion of the pond in our stewardship taught us about ecological justice and the need to lead by listening and, if necessary, by changing plans.

In this way, in making gentle decisions that would protect the integrity of nature, the students began to love the land. They learned that teaching similarly requires expanding notions of care. After 'falling in love' with the abandoned piece of land covered in weeds and debris, they fell in love with the pond. One day I arrived at the garden to find my students engaged in a difficult operation of removing beer cans, garbage and an old rubber tire from the middle of the pond. They were standing on a very narrow log, holding a big stick. Class had not even started yet.

> In those days we lived beside the river. An utterly beautiful river, with its depth turned in our direction. My grandmother said that the river was a baby bottle for the whales and their children, formula which went to their heads and made them sing and snort out loud. The whales lived further upriver, rather far from us, but on beautiful summer evenings, if one used one's ears and a little imagination...a murmur in the air...a great stirring of the depths...a monumental door swinging on its hinges...there was a whale approaching. (Marchessault 1990, p.176)

Where to Begin: Ecological Design

> Before you disturb the system in any way, watch how it behaves...starting with how the behaviour of the system forces you to focus on facts, not theories. (Meadows 2005, p.196)

Besides the decision to leave the pond alone, the students developed their garden design plans, another eco-centred activity. The assignment began with a few simple instructions: design a garden that included raised beds and a place for students to gather; we would then share the designs.

The student designs were placed on a screen in the classroom and included a mix of hand-drawn symbols, squares and circles, combined with computer-generated garden designs. One design clearly stood out: it was irregularly shaped, with the exterior parameter of the garden bulging into an oddly shaped arc. After a moment, we realized that this (crude looking) design was in the actual, irregular shape of the land itself. The plan included areas for garden beds which led out from a (natural, tree-shaded) classroom area to composter and soil bins. The plan was organic, irregular and fit the imperfectly shaped land perfectly.

The assignment taught us how eco-centred design methods begin with the land instead of with preconceived, abstract images. Our experience at the site—weeding, looking around and listening—contributed to our understanding of an effective, organic design plan. In taking our time in deciding on a design plan, in spending time at the site, we could now actually visualize the garden. A row of pine trees on one side of the classroom would provide shade during the warm Okanagan summer months; heat-tolerant aromatic herbs (i.e., rosemary) would be planted above the pond in 'the sensory garden'. Spending time at the site, we also became aware that since the garden is located in a semi-circle of trees, the shade and light is constantly changing. Now we knew the importance of the pond.

In the design we eventually chose, the 'classroom floor' faced east, towards the rising sun. We decided that the boxes could be laid out like sun rays, shining in a westerly direction, towards the pond. (One year later, a student flew over the garden in a light plane and noted how the shape of the garden was, interestingly, the same shape as the pond.) The students were learning to work with and design the land, by listening to the land.

Building an outdoor learning site for learning *about* sustainability continually tests the principles of sustainability itself. In the building process, goals change and sustainable learning begins. In building Meadowcreek, a model outdoor learning centre, David Orr writes how nature similarly 'interrupted' their plans with intense floods and drought. He writes about his fixation on trying to 'neaten' the edges of the land when, obviously, "land has a mind of its own to which we are not privy" (Orr 2005c, p.97). As the student teachers and I similarly discovered in the garden, teaching that begins with outcomes can preclude an organic element from emerging in a lesson; learning takes place in the interchangeable dynamic between student, teacher, community, context and unknown elements of nature. Learning about sustainability within the land allows room for 'the unknown' to emerge; in ecological design, the environment shapes knowledge. In deciding a 'curriculum' in and for land-based learning, student teachers are thus exposed to more concepts of effective teaching: spontaneity and improvisation.

> Unlike industrial design, ecological design relies on the interaction between humans and land, or 'the careful meshing of human purposes with the larger patterns and flows of the natural world...and the study of those patterns and flows to inform human purposes'. (Orr 2004, p.104)

Common characteristics of ecological design include

- right scale;
- simplicity;
- efficient use of resources;
- a close fit between means and ends;
- durability;
- redundancy and
- resilience. (p.105)

'Outcomes' of 'good ecological design' promote

- human competence instead of addiction and dependence;
- efficient and frugal use of resources;
- sound regional economies;
- social resilience. (p.105)

Along with design principles set out by environmental thinkers before us, we learned that ecological design also begins by going outside; it begins in the going. Many classrooms are windowless and situated in the interior blocks of large concrete buildings. The message in much school design speaks loudly: do not (literally) think outside of the box. In fact, a great deal of effort is spent keeping students inside, away from windows. Is it any wonder that students will approach learning with a severed view of nature, that their first encounters with nature will inevitably turn towards depletion of natural resources in ways that reproduce a disconnected view towards the earth?

The students at the Learning Garden began with computerized design models and notions about how to use water in the pond. Through hands-on work, students learned the first principle of learning in the land, which would also perhaps shape how they approached teaching: first, they would *do no harm*. Their work quickly became about nurturing nature, which began to feel sacred. The students thought more about future stewards. Would others learn the bond that had started with this first group of students? One of the recurring themes in student journals towards the end of the course was how the site would be treated by future students. As one student wrote:

> My reservation about the garden...is that we will do all this work, invest all this time, energy and emotion into the project only to see it falter after we're gone...I have become very comfortable on the garden site and know every little nook and crannie, every rock and every mound of dirt in the area and I find that I have become quite attached to the project and the worksite. So...it is going to be hard to say goodbye to it.

A theme that would be repeated during the first four cohorts was that once the students gained firsthand experience of the land, they saw their work in a larger process in both time and space. In the garden, 'the environment' became both object and subject of the learning process as a site that constantly promoted the idea of learning as process in an interchange of ideas, community and practice. Such notions are uncommon in most academic campuses, where usually the actual physical place "is intended...to be convenient, efficient, or aesthetically pleasing, but not instructional. It neither requires nor facilitates competence or mindfulness" (Orr 1999, p.229).

A lack of direct experience with 'place' in higher learning had been the experience of most of the students prior to their experience in the Learning Garden. While the garden began to teach the students, I continued to wonder how mainstream schooling and teacher education had for so long actively excluded the land, even as school gardens had thrived at various points in the past two centuries. I recalled the words of a teacher educator colleague I had met during my graduate student years: "I don't take my students outside. Because I am modeling *not* taking students outside." How had Western teacher education sustained a Victorian attitude with regard to the separation of humans from nature?

In indigenous worldviews and traditions of learning, 'place' is central to teaching and learning; people are encouraged to listen to and learn from the land. In such a context, what happens when teaching and learning occurs without any connection to place? A Yupiaq researcher from Alaska notes that for many indigenous students around the world, formal learning exists in an "alien school culture" where "curricula, teaching methodologies, and often the teacher training associated with schooling are based on a world view that does not always recognize or appreciate indigenous notions of an interdependent universe and the importance of place in their societies" (Kawagley & Barnhardt 1999, p.117). The same researchers also note how learning that emphasizes "local knowledge" that "originated under conditions of marginalization, have gravitated to the centre of industrial societies" and that, more recently, "the pedagogical solutions that are emerging in indigenous societies may be of equal benefit of the broader education community" (p.121).

As 'pedagogy of place' in indigenous societies is passed through generations, as stories are told about resourceful, sustainable ways of working with the land, place becomes a holistic theme in learning; 'place' is survival and wisdom. In traditional societies, the land unifies learning. When learning begins with direct experience of nature, the idea of compartmentalized knowledge (i.e., knowledge separated by academic disciplines) is useless unless it contributes to a unifying concept of nature. In other words, when land is directly integrated into the goals, theories and practices of learning, the evaluation of learning and knowledge also shifts, from learner to the land.

In a garden, in teacher education, indigenous and Western systems can be integrated around themes of ecological design that places the land at the centre of learning. If our garden were an example, learning could actually be centred on the pond, due to its local and global significance, and could include discussions on water's biotic and sacred qualities; the value and moral obligation of providing habitat for migratory species; the micro-biotic quality of the water; the weights and measures of water saving potential; conceptions of indigenous and Western ideologies; and the science of water conservation. Such a blend of worldviews, in an interdisciplinary context, could help open discussion and address issues in social and ecological justice. In this way a garden provides a context for ecological learning design inclusive of place, knowledge systems and culture.

Conclusion: The Land Connection

> We give away our thanks to the earth
> which gives us our home.
> Dolores La Chapelle (cited in Roberts & Amidon 1991, p.239)

Firsthand knowledge, and an openness to interdisciplinary ideas, began to influence learning and decision making in the garden. Such learning, especially in teacher education, was difficult to place inside any previously established learning model. Most of the students were practical-minded teachers, or on their way to becoming teachers. While most launched into the garden with the idea of 'getting a job done', in time, after having made their first context-based decisions, something extraordinary took place. Teachers who had been used to working with 'efficiency' uppermost in their thoughts began to open to new ways of learning. Here they were, digging and learning in a garden, with their hearts and their minds, without a direct sense of personal gain. It was, in many ways, the perfect metaphor for training new teachers.

> Traditional societies personally transmit and personally use knowledge...Knowledge exists only because one person gave it to another, and it is kept alive only by repeated use and personal transmission. (Margolin 2005, p.73)

Amidst the heat and stinging weeds, the students thought about future learners. Part of the wonder of the garden was that, in the end, it was not their own. It was a gift. It was a valuable lesson in learning how to teach well. For the students who would move on, graduate and teach, the sense of limited time on the land brought profound connections. The garden is, like teaching, a labour of love.

A student in the first cohort described the garden as a metaphor for restoring the five senses:

> Having spent time in the Learning Garden—from its initial conception to its early construction—I would have to agree that my five senses were restored...And I hope that the garden's legacy will be one that allows all who visit to undergo a sort of restoration.

The garden was also a gift that would hopefully permit future students of the learning gardeners to experience the same kind of restoration:

> It is impossible for a child to work [in a] garden without tuning himself to certain universal laws...while they are grubbing in the earth, stirring the soil untiringly so as to let in the moisture and the air, nature's secrets are sinking deep into their heart. (Dora Williams, from *Gardens and Their Meaning* 1911. Cited in Welby 2003, p.2)

At first the garden was a backdrop for ideas of land cultivation and academic learning. In time, the idea of a garden as 'land' began to blend equally with the realization that garden labour is linked to social and ecological justice and community. As they weeded, worked and discussed, students gradually began listening more closely to the context of the work: the birds, soil, water and shifting earth and sun. Through mindful interaction, the garden, as environment, kept returning into the foreground of learning. After a while it seemed impractical not to make group decisions based on the needs and changing rhythms of the land. Students took up their visions, and adapted those visions, for the garden; they worked in the intense summer heat, and by summer's end, they were filled with a sense of accomplishment looking back on the process itself. There was nothing left to do but give up their work, and their labour, as gifts to unknown, future stewards.

The Learning Garden was neither a 'science nor a scenic' project. As an outdoor classroom, learning in the garden moves critical theory to practice, enabling 'hands on/minds on learning' and helps realize Orr's (2004) idea of teaching through the "concrete reality of natural objects" (p.96). From concept to infrastructure, students discovered how, in a garden, ecological-centred practice and philosophy go hand in hand. It was not lost on students that a place could so dramatically influence plans, discussion, the way learning took place and the growing respect they felt for the environment and for each other. In the garden, 'the environment', the land itself, made the interchange between theory and practice real.

We began with the garden, and somehow, the garden changed long-held conceptions of academic success based solely on a model that prizes efficiency, speed and individual success. In the garden, there is constant flow between practice and concept, tangible and intangible ways of learning. The garden also includes the paths surrounding the garden and the stories told. The pond is water; it is sacred and it is also a site that brings learning, conservation and (later) activism. Since we decided not to use the pond, the pond provided us with a reminder of core environmental ethics related to the need to find alternative solutions to our needs. Ultimately, the focus of the garden is not our lessons, or philosophies, or even our ethics, or even our cultivation techniques or even the concept of sustainability. Sustainability conceptualized in the garden focuses on ecological justice because the needs of the garden are always considered in the larger context of learning and life cycles, of which we are a part. The process of learning in the garden always begins with, and returns to, garden and pond, land and water.

UN scientists' hope for human adaptive capabilities in dealing with climate change includes the work of farmers. The wisdom of those who work with the land can teach teachers how to teach.

> Each year is like starting over: the climate is different; the marketplace has changed; the condition of the soil may have improved but in subtle and unpredictable ways. (Ableman 2005, p.177)

Uniquely, student teachers arrived at their goals for learning through working the land. One of our larger garden goals was to realize, through hard work, and without a lot of fanfare, David Orr's (2004) concept of ecological design intelligence, or "the capacity to understand the ecological context in which humans live, to recognize limits, and to get the scale of things right" (p.2). We could apply this idea to the garden by making context-based decisions, instead of randomly applying abstract ideas from, say, a textbook or garden guide. In fact, our work in the Learning Garden allowed spontaneity and mistakes to influence our learning process. In working through the challenges, we tried, like Orr, to develop an eco-centred model of success in learning to approach land "carefully, lovingly, competently" (p.105) in ways that include firsthand, practical knowledge. We learned how love for the land, the pond and our larger learning environment was practical, as it led to deep concern, care and thoughtful actions that would help the natural area sustain itself.

Working with pre-service teachers in a school garden provides vivid metaphors about the nature of teaching and learning itself. The garden began in an unwanted, out of the way piece of land covered in weeds and unwanted piles of metal, glass and other debris. Such lessons help students learn the value of 'blooming where they are planted', in less than utopian conditions, which might also apply to future teaching situations. In this way, the garden helps new teachers face teaching with a sense of possibility, idealism and courage, the same way an inexperienced, brand new teacher learns to face hundreds of students each day. As the farmer in the class said on the first day, hope is practical.

What is the role of a garden in teaching students about eco-centred learning? Building and planning the garden during the first cohort and planting, replanting and harvesting during subsequent cohorts emphasized the process and cycles of teaching and learning. Nurturing plants from seedlings, observing their growth, as students and teachers learn from the garden, is a powerful way to help future teachers learn how to learn. In later cohorts, initial reluctance gave way when students worked together to apply their knowledge. If, for example, a student wants to plant a rose, instead of native, drought-tolerant plants, it is obvious that a rose in the local climate would need a lot of water and care. Is the student willing to provide that? Is a rose practical in a desert landscape? What are the cultural assumptions that lead the students to believe a rose is 'beautiful' because it uses 100% more water than a native plant such as the Oregon grape? For students new to a garden, learning does not lie in certainty, or in applying previous ways of learning, but in a willingness to learn and adapt to changing situations.

With regard to visions and ideals about the land and learning, the student teachers and I grew alongside the garden: unpredictably, in the context of organic life. A garden reveals how the process of learning, rooted in the context of one's surroundings and community, becomes the lesson itself. To learn in a garden with students is to be in a constant state of environmental and social activism. As veteran social activist Grace Lee Boggs states, a community garden is a sign of "hope trumping social despair" at the grass-roots level, where we "regain our humanity in practical ways" (cited in Moyers 2007).

CHAPTER FOUR

✿

Garden as Community

Introduction

> It's not all sunshine and daisies. (Tracey 2007, p.2)

At the end of summer, the Learning Garden was passed to the next group of stewards, the middle school teacher education cohort. In the garden as *community*, student teachers learned the importance of respectful interaction with their location and with each other.

In turning towards community, and promoting values of equality, the garden was influenced by local Okanagan learning traditions. The ideas in TEKW support firsthand practice and cultural knowledge around native plants as a principle of sustainability and as a way of life (Turner, Ignace & Ignace 2000). The tradition of First Nations is to approach the land with an attitude of gratefulness and to view the earth, with its capacity to produce food, water and survival, as an equal partner in community.

> The origin of the Syilx is the Earth. We were created from the land as part of the land, therefore the Earth in all its elements and the ones who came before us, are our 'parents'. Like all 'parents' their job is to take care of us and as such everything we have was given to us by them. We never 'accidentally' come across fire, it was given to us, food-given to us, dreams-given to us, everything was a gift to ensure our survival and they were all given to us, by our 'parents'. This is what we mean when we say 'the living Earth' and this is not a term to be thrown around lightly. (Syilx Nation Okanagan First Peoples 2008)

Can such ideas influence mainstream teacher education and help student teachers take up both the practice and idea of land *as community*? Can community become a core process in learning how to teach and learning how to learn? As described earlier in this book, one small campus garden can help students understand the practicalities of sustainable land use and perhaps even shift behaviour; can it also influence attitudes towards teaching and learning itself? In the narrative of the garden, the move towards inclusivity and community was a dramatic shift that helped students expand their understanding for both ecological and social justice.

The first cohort of students intentionally shifted 'place' to the centre of their work. In time, the concept of place began to influence, and direct, their actions and their learning process. As the garden grew, the students learned more and more in/as community in both practice and spirit. Over time, as 'place' itself became the centre of decision making and class work, 'community' (local, global, social and biotic) helped create deeper changes in thinking. It became obvious that the Learning Garden's existence as a site for environmental learning, and as a model of progressive learning practice, depended on students' ability to plan, work and learn both independently and as a group.

At the same time, 'deeper' shifts in attitudes took place; even if students worked the physical garden to 'perfection', if their work remained in a model without shared ideals of community, the garden as a site for deeper ecological understanding would wither. In the next cohort, we learned how a campus garden can include community in/as a deeper learning process. In this way, the garden, both as physical space and as conceptual community model, challenged the roots of academic training. During the next cohort the garden began to evolve into a *Learning* Garden.

What Happened?

> What are the dangers of education? There are three that are particularly consequential for the way we live on the earth: (1) that formal education will cause students to worry about how to make a living before they know who they are; (2) that it will render students narrow technicians who are morally sterile; and (3) that it will deaden their sense of wonder for the created world. (Orr 2004, p.24)

Instead of prizing 'ownership' of land or ideas, the Learning Garden began with an ideal of shared local knowledge of the land, which was given as a gift from previous learners to future learners. The garden stewardship was thus conceptualized as a community model, where the 'current' cohort's main responsibility was not so much to 'make their mark' but to provide a bridge between learners so that the land could sustain itself. The first cohort of students in the garden had all enrolled in environmental education and thus shared many of the same, unspoken values related to ecology and community. The first group's gift to future stewards was based on ideals around leaving a legacy of physical space and of community that would be 'received' by students in teacher education.

From the perspective of indigenous teaching and learning (Kuokkanen 2007), the process of giving the garden, and sharing knowledge, has larger implications for sustainability and learning. As Sámi scholar Kuokkanen says, "For indigenous people, the world's stability, its social order, is established and maintained mainly through giving gifts and recognizing the gifts of others, including the land" (p.7). Kuokkanen explains the 'logic of the gift' in indigenous systems of knowledge:

> Instead of viewing the gift as a form of exchange or as having only an economic function, I argue that the gift is a reflection of a particular worldview, one characterized by the perception that the natural environment is a living entity which gives its gifts and abundance to people provided that they observe certain responsibility and provided that those people treat it with respect and gratitude. (p.32)

On the idea of the earth as a larger, living social and ecological network, she adds: "Social ties apply to everyone and everything, including the land, which is considered a living conscious entity" (p.33).

That the garden embodies a gift, with new stewards arriving each year to continue the legacy of giving, and re-giving, is an ideal that helps describe the teaching process to new teachers. An ideal of knowledge as 'gift', with an emphasis on the transaction between students and the land, sustains the idea of learning in/as a community process.

Brand new student teachers, many with freshly completed undergraduate degrees with a major in one or two subject areas, tend to enter teacher education with fixed notions of teaching and learning. Many of these students want a quick, practical route to becoming teachers. Success in the education programme, and as a teacher, means relying on patterns of learning focused on competition and individual forms of achievement. The prospect of learning with an emphasis on learning, on giving, on community, instead of on individual ways of knowing, challenges a worldview where they have found success (and high GPAs [Grade Point Average]).

When I told the next group of student teachers they had the option of developing curriculum and methods around ecology and a garden, a handful seemed interested. Some wanted to know if this was 'the norm' in schooling. Student questions lead to discussions around leadership: What is the role of a teacher in a school? What is a teacher's responsibility in helping students value the earth? In informal class discussion most students agreed that 'helping the planet' was obvious, urgent and needed to be a part of their professional training. Some students also saw a moral imperative in teaching students 'about the planet'.

The challenge of teaching student teachers in a garden is not whether they will add 'eco-centred lessons to their teaching, but whether they are willing to adapt to their roles as teachers who give voice to the land and community; the challenge is whether they are able to move towards 'learning to teach' by learning to learn. The shift towards learning in/as a community does not offer a guarantee of individual academic success. Teaching through a garden, and through a community, is thus, for many, a challenge to their preconceived notions of academic success and to their imagined role as 'teacher'. In eco-centred teacher education, even students who are not yet teachers must re-imagine their roles.

In an ecological approach to teacher education, one that includes ecology and community, students learn how to learn. Such an approach to teacher education can provide a model for learning where *unknown* knowledge, knowledge of nature and community, is equally valued.

After the first day, one student told me: *I hate nature.*

When Student Teachers Encounter Ecology

Bowers (1999) describes the controls of traditional teacher education:

> many students enrolled in teacher education programs bring another con-cern that makes them even more resistant to rethinking the connections be-tween the cultural assumptions underlying both the explicit and implicit curriculum of the current ecological crisis. This concern is based on a deep fear of losing control of the classroom, which leads to their feeling that dis-cussions of how the language of a culture reproduce...are unrelated to their need to learn teaching and behavioural management strategies. This fear of the liminality that is part of every teaching/learning situation adds a further level of difficulty in reforming teacher education in ways that take account of the culture/ecological crisis connection. (Bowers 1999, pp.162–163)

The students had been trained in their undergraduate education in all the "modern assumptions" about learning, which include "viewing change as progressive in nature, intelligence and creativity as attributes of the autonomous individual, science and technology as the source of empowerment, and the commoditization of all areas of community life as the highest expression of human development" (pp.162–163). Such commoditization and ingrained bias provided a constant series of barriers to developing a learning community in the campus garden. The notion of working on a shared parcel of land, for the benefit of the larger community, challenged their previous assumptions about land, learning and 'successful' academic progress.

Reclaiming the commons requires taking back the ability to make decisions collectively and reflectively. For teacher education, this allows open questioning, as Shiva (2005) states, "Relationships create the space for people to respond to one another. This creates responsibility and establishes the ground for sharing and compassion" (p.89). The learning processes of the garden posed a direct challenge to the student teachers' preconceived ideas that supported the 'commoditization' of 'community life' since the entire course was centred around the development of community as a means of tangibly supporting ecological and social justice. Simply put, the goal of the course (and, therefore, the standard for evaluation) was to (1) create garden projects that helped nature sustain itself and (2) use the products of that work to support the community. Ironically, even in the goal-driven, success-oriented context of traditional teacher training, a community garden without a community could never be a 'success'. Working towards an inclusive community in the garden thus redefined academic success in very real ways. What cultural values are at work when students resist learning about their environment?

> Unfortunately, the power of the media, the individually centered and industrially oriented nature of the school curriculum, and the constant encounter with peers who have been indoctrinated with the values that equate social status with brand name consumerism, compete with the values and way of thinking reinforced by participating in community gardens. In too many instances, the influence of industrial culture has the effect of marginalizing the experience of gardening or storytelling as being special, but not part of the mainstream of daily life. (Martusewicz 2006, p.53)

The sharing of practical knowledge—engaging in physical outdoor work, within traditions of community—challenges entrenched views of mainstream consumerism. In fact, the first impression of a garden is often antithetical to what students have been trained to consider as signs of 'value' and 'success': the garden seems small, the work involves (often lacklustre) hands-on labour and the (intrinsic) rewards seem invisible to the gaze of mainstream socio-economic power structures. Why would a student take interest in it? Why would a student want to learn how to learn when they can *know* within a proven model? Even the students' initial reluctance provided an important lesson for student teachers: invisible, intrinsic rewards in small-scale, non-corporate settings largely describe the work of most school teachers. And, like school teaching, the work also requires a leap of faith.

As with later cohorts, as students began working in the garden, their objections largely dissolved as they realized that learning in a garden requires a new set of expectations around care and community that are not impossible to understand, but nonetheless often require a break from previous models of academic training. Initial, individual reluctance usually gives way to enthusiasm, idealism and commitment within community. For most, it is more the idea, the external, conditioned impression that learning through principles of ecology, in a garden, is *just too different* from their preconditioned ideas about teaching and learning. Within their experience of formal education, most student teachers have rarely learned within the framework of commitment to the land through positive, powerful community. Most have never stepped outside during their university classes. However, the experience of hands-on learning, the practical work, also guided students towards positive ideas about 'place' and 'community' where they could begin to imagine their work. Going to the garden, and allowing learning to be influenced by the setting, helped change minds.

In traditional indigenous education, learners 'learn by doing'. Since I started working with students in the garden, learning by doing has taught me about teaching teachers. Experiential learning involves hands-on experience; in order for students to get to that experience, educators must help students cross ingrained, ideological or perhaps even imaginary borders, including the bias that community, manual labour or farming are less valuable than abstract information taught in a highly controlled setting of the classroom and mainstream academic structures.

Stumbling towards Community

Some students told me that they would much rather 'work independently' than in a community setting. Some did not want to leave the classroom with its inherent hierarchies. For these students, the practical knowledge needed to be involved in a garden seemed too great, the learning curve too steep. I sensed that to learn new knowledge, publicly, within community, would be a kind of embarrassment. Some would rather learn on their own, after which time, presumably as teachers, they were meant to return to the community and act as leaders. Across many garden cohorts of both student and practicing teachers, I have noticed a kind of fear of embarrassment, a fear of 'not knowing', even as students are invited to learn creatively, in community. Student teachers are trained to want to lead, instead of learn; as the newest members of the profession, they are afraid of making mistakes or of looking as if they do not have all of the answers. Such an attitude has long influenced mainstream ideas of what it means to 'be a teacher'.

And now, the planet is changing. Teachers and students must work together to solve problems for the planet in ways they might be used to; a fear of 'not knowing' is in fact a major obstacle to open-ended community discussion that includes a much needed acknowledgment on human limits. Paradoxically, it is in sharing in community, both as individuals and as group members, that knowledge expands. I have seen time and again how, after students take the first step towards participation in the garden community, they find that they have valuable knowledge to share. As knowledge is shared, understanding grows; their understanding of learning, community and knowledge also expands. In working in the garden, I see how the first step towards teachers and students learning in inclusive community also involves the inclusion of oneself in that community. The circumstances of their previous mainstream learning (individual work, ingrained notions of teacher-learner hierarchies, emphasis on abstract and product-oriented models, the student as 'star' candidate, students learning without the natural environment to guide them) contribute to their hesitation to involve themselves in a learning environment that includes shared learning processes, where the process of learning becomes part of the learning. To learn how to learn, for those conditioned by industrial models, means risk. In a planet in crisis, to continue to learn in an industrial model, without questioning the underlying assumptions that guide mainstream Western knowledge processes, is a greater risk.

In a garden, the idea of a single, individual student contributing the basis of all knowledge is not a sustainable notion of stewardship, teaching or learning. Instead, in a garden, pathways to larger understanding begin with sharing and with the actual *doing*. Amidst the tensions, the stewards and I began to understand how a *Learning* Garden requires hands-on work, and perhaps more importantly, it requires a conceptual philosophy, based on shared principles and a communal work ethic to guide knowledge, labour and inspiration. Learning in a garden, in a shared, community model and value system, is also an evolution within the process of that ideal. Where I had thought the greatest threat to including ecology in mainstream teacher education would be lack of student interest, I realized that the real clash takes place when old processes of learning meet in a new context for learning (i.e., in a garden). The (often hidden, unknown) challenges posed by a garden learning community to previous notions of teaching and learning is, by far, the most important, and difficult, learning experience in a garden. It was a tension that needed to be discussed, welcomed and written about. The group would have to unlearn bias towards 'tension' itself. As Ted Aoki (1991) suggests, "In our North American tradition, we tend to appropriate tension negatively. We seem to succumb to an urging within us to reduce it or even eliminate it" (p.183).

Tensions and 'mistakes' in a garden are welcome. In a *Learning* Garden, 'mistakes', when they are realized and brought forward, are the live examples by which learners (and teachers) learn to learn. For 'success-oriented' students, the garden allows them to experience an imperfect, incomplete, organic learning cycle that includes both their own mistakes and acknowledgment of the imperfection of learning and teaching itself. As a teacher educator in a mainstream classroom setting, I would never have been able to effectively teach student teachers that experience; community, *mistakes* and *problems* are part of a 'successful' learning process. The lessons of becoming a teacher, learning how to engage others in unfamiliar processes of community, requires openness, experience and, perhaps more importantly, involves allowing mistakes to inform the process. The Learning Garden taught teachers that the more mistakes they 'allowed' themselves, the further they moved away from old patterns of thinking which rewarded the external attainment of knowledge isolated from social and ecological community. Working through tension, then, became another part of the learning to learn process.

Other tensions arose that proved to challenge previous understandings. It was difficult to collectively decide on 'what to do' with the garden. The ideals of the garden as a cooperative, shared model of learning also made us aware of land and learning models based on ownership and personal profit. The cohort of student teachers exemplified the challenges of developing a community model of learning within pre-existing models. During that year, threats to the garden community (physical, ethical, external, internal) all somehow related to concepts of individual ownership. In a Western model of education, it seems that just as people care about land, they also want to control it. However, the second cohort, in their difference, within tensions, provoked complex, firsthand understanding of the connection between ecology and community. While the first cohort provided an example of community with like-minded vision, the second cohort provided even more evidence of the need for learning how to learn within a community, and a process, which gives voice to the land. The second cohort also revealed the dramatic changes that take place when ingrained attitudes are challenged.

The lessons of the second cohort brought forward a practical need to review some of the original visions of the garden. A garden becomes sustainable through inclusive involvement with, for example, a rotating stewardship model where each cohort collectively decides on a project each season, in the context of local, current ecological and social concerns. A sustainable garden requires a combination of listening, planning and allowing community to 'take place'. Such learning also requires reflections on the ways the practical work is tied directly to values, ideals and visions around teaching, learning, culture and community. In short, a Learning Garden involves both practical work and shared discussion around theory, practice, teaching and learning.

In the case of our garden, the need for deeper connections was prompted by tensions, as previous methods of academic learning met unfamiliar ones. The need for community and shared decision making, the idea of learning how to sensitively and respectfully speak for the land and with one another, also brought potential for transformation.

A learning community is not a 'natural' state for those conditioned to teach and learn in isolation. Alternately, students used to a community model of learning, learning within a hierarchy of solo achievers, can be equally uncomfortable. For many, an eco-centred community model of teacher education is at striking odds with a college system biased towards individual stamina and success. To develop closer connections between students and the land, for schooling to become truly sustainable, educators must protect and promote ideals of community learning to students trained in academia who might not have a single prior experience of a 'learning community'.

Ultimately, the idea that a sustainable planet depends on practical, hands-on knowledge gained ground. The students accepted the imperative that ecology needed to be a central part of their training. The garden transformed student teacher reluctance into deeper, hopeful awareness. Tension in the garden brought new directions and redirections.

Saving the Pond

> Stewardship is an ethic and practice to carefully and responsibly manage natural resources and ecosystems for the benefit of current and future generations. Stewardship demonstrates a commitment by governments, communities, corporations, non-profits and individuals to voluntarily act in an environmentally, socially and economically sustainable manner. (Stewardship Centre for British Columbia 2007, p.4)

In spring, a small, dedicated group within the teacher education cohort, post-practicum, continued building the infrastructure of the garden by building up the soil and designing the beds. We then learned of a campus plan to drain the pond to make way for a new building. Just a few months before, we had begun work on a weed-and-junk-filled, largely 'hidden' part of the campus. The garden, surrounded by development and roads, located at 'the back' of the campus was suddenly at the front of the campus. All of a sudden, the garden, the natural environment that surrounded the classrooms on campus, was 'in the way'. When I told the students about the new building, they wondered how the goal of sustainability fit with the pond removal. One argument for removing the pond was that the pond was man-made and, somehow, 'unnatural'. A discussion took place beside the pond one day:

> Student A: What are the ethics in creating a natural system, that turns into a complex ecosystem, then changing our minds about it? Just because we change our minds doesn't mean it's no longer an ecosystem. We built the garden because the pond was here. It's part of the environment, you can't separate it...there's not a *here* and a *there*. How can one place be two places?

> Student B: It's a watershed; it's still natural no matter who 'made' it. Those trees and reeds and rushes are natural, whether or not we're using them for anything. Does it become ours when we decide to use it? If the pond is 'not natural' try telling that to the migratory ducks who live there.

Questions around the pond found the stewards taking responsibility for campus decisions that would have a lasting impact on their surroundings. The previous lesson in making positive, conscious decisions for the garden (i.e., leaving the pond alone) taught students the importance of listening to all members in decision making. The land taught the stewards to listen. This time, quite naturally, a small group of student teachers found themselves activated around the environment. The group discovered values which focused and guided their work. The students clarified their involvement in the garden; their actions were an extension of their ideals, their ideals an extension of their community. Student teachers had thus evolved from university students enrolled in a professional program to land stewards guided by ethical principles. In responding to the needs of land and community, their work was similar to Wendell Berry's (2002) vision where a community "must change in response to its own changing needs and local circumstance, not in response to motives, powers, or fashions coming from elsewhere" (p.163).

As a relatively new professor, I was not certain I wanted to make 'waves' in the pond. But the spontaneous action of providing stewardship for the garden, then the pond, was based on values that I, not so much as an activist, but as a teacher educator, strived to model for my students. I felt a social, moral, cultural, community and professional obligation to promote, even defend, the pond. The garden was, after all, our classroom, and a classroom for future learners.

Supporting the land through ideals of a minority within the context of conflicting larger powers is, I learned, how true land stewardship starts. While building the garden infrastructure in the first cohort, the students developed a deep bond with the place as they passed on the gift of their labour to future students. In the following cohort, students enrolled in a professional program, whose work in the garden was largely voluntary, ultimately responded with a deep and abiding interest in ecological justice, prompted by love of the land and community. When they thought of the hard work of the last cohort, and realized they were a bridge to future stewards, the decision to support the pond was easy.

One student offered to live in a raft on the pond in order to save it from destruction. During class he told the others of his new plans. "I can make breakfast out there!" What could I say? To invite students to participate in learning through the land is an ongoing commitment that must take place in both discussion and, ultimately, in action. When students give voice to the land they want to continue speaking.

In the midst of campus construction and land development, we had a discussion one day about land ethics and the importance of preserving and defending biotic life within the larger life cycle:

Student A: Selling and paving agricultural land...it will never come back. At what point is human action on the earth—ecologically wrong? When it becomes inorganic? At what point have you gone too far?

Student B: Recycling is genius. You have to preserve the process...if you respect life—you respect death. When you remove death you remove life. When have humans gone too far? Maybe...when life is so contained in such a way and on a scale where it can't come back to its natural state.

It was during the second cohort that the ethics and principles, the 'soul' of the Learning Garden, took root. The garden was not an abstract 'vision' or a 'critical model' of successful sustainability, but an actual place where learners experienced, firsthand, community-based decision making around on a true land ethic. In this sense, a goal of sustainability had to be tempered with the ethic of leaving the lightest footprint possible. Students recognized the 'genius' in the cycles of life, death and renewal and realized it was a sacred value created by the earth that was worth protecting. Sustainability now included limits on human impact, as the students understood that any form of human intervention could bring the land past the point of returning to an organic state. We learned that the notion of impermanence, with emphasis on natural, organic processes, was an ideal state for both the land and learning. The idea, and the practice, of teaching and learning as a 'gift', in the way that the first cohort had laboured for future learners, was part of honouring that impermanence.

The students now saw the dangers in claiming individual ownership of land and ideas, as such a view would severely limit the legacy, the dreams and the ideals of future learners. The garden was teaching teachers that learning meant shared interaction based on mutual relationships of trust and care; they learned how students, as a life force, must learn beyond teacher controls. They learned how teachers must also step out of the notions of control in order to free their students. If the garden was a classroom, it was the role of the teacher to make room for 'crops' which would permit sustained growth. All of these lessons arrived from the physical space of the garden, but the garden now grew into an ideal, tangible model for learning. The students provided stewardship for the garden, for the surrounding area (i.e., the pond) and for the ideals that grew, and would grow, from the garden.

In the garden, the student teachers moved beyond their previous models of learning in gaining understanding of a garden as environment and as community. From the first invitation to learn in the garden, the students began to re-imagine their perceptions of what it means to 'teach' and to 'be a teacher'. The idea of re-imagining the role of teacher does not arrive naturally in an education system where the notion of 'good' teaching is often associated with high levels of control exercised by 'powerful' individuals who make a 'permanent' difference. The garden taught us that a 'powerful, permanent' impact, even one fuelled by the best of intentions, could bring an organic compound past the point of renewal. To teach sustainably, then, means allowing students to make mistakes and 'return' to a point of not knowing, to renew themselves in an organic process.

What would the world be, once bereft
Of wet and of wildness? Let them be left,
O let them be left, wildness and wet;
Long live the weeds and the wilderness yet.
– Gerard Manley Hopkins, from *Inversnaid*.
(Cited in Roberts & Amidon 1991, p.154)

Aiming towards Wholeness

> [M]any Native as well as non-Native people are recognizing the limitations of the Western educational system, and new approaches are being devised.
> (Kawagley & Barnhardt 1999, p.117)

The evolution of the garden raised ecological consciousness and brought new questions for teaching and learning. The language in education is often about defining 'problems' and posing 'solutions' through learning 'outcomes'. From an ecological perspective, David Orr (2004) talks about the 'problem' of education itself, how humans, themselves, pose 'a problem'. As humans harm their environments in the quest for knowledge, they harm the shared space, without even knowing the results of their actions. Orr defines this as a symptom of a society that does not necessarily require more 'knowledge', but desperately requires the values needed to guide that knowledge. He says "in the confusion of data with knowledge is a deeper mistake that learning will make us better people" (p.10).

New writings in ecology challenges teacher education to re-examine a standard of training teachers 'just enough' to teach in the classroom. Such teachers will presumably go on to train their own students 'just enough' to enter the workforce. However, as Orr's comments imply, economic changes brought by climate change will also change the workforce. New ecologically conscious teacher education also raises new questions: Why continue to teach students in ways ill-suited to a planet in change? Should we not prepare teachers to work within processes and communities of change?

Orr's (2004) imperative is that humans must move away from the notion of 'all-knowingness'; instead "we must pay full and close attention to the ecological conditions and prerequisites by which we live. That we seldom know how human actions affect ecosystems or the biosphere gives us every reason to act with informed precaution" (p.xiii). His views on learning and intelligence carry deeper implications for teacher education: "True intelligence is long range and aims toward wholeness. Cleverness is mostly short range and tends to break reality into bits and pieces" (p.11). Thus, in responding to Orr, and to the deeper concepts of environmental philosophy that must now guide our work in classrooms, how can teacher education evolve in a way that 'aims toward wholeness'? How should we educate students towards ecological 'intelligence' in a changing world and workplace?

Connecting students to an outdoor space can provide the necessary conditions for students to think 'long range' and 'towards wholeness'. A garden is one way to help students redefine both 'intelligence' and academic 'success' by providing a new context for learning. Evaluating success in such a community also means acceptance of difference of all those who want to contribute. Learning through land stewardship also means being able to distinguish between 'long range' goals for continued learning and sustainability and 'short-term' involvement. As the garden evolved, I began to understand what David Orr meant by 'the end of education'. Part of learning through environment also involves moving away from harmful practice in unfair circumstances, even though such practice appears dynamic and outwardly 'powerful' in the short term. As Orr (1992) says, "students need opportunities to work together, to create, to take responsibility, and to lead in a community setting without which they are unlikely to comprehend the full meaning of virtue, ecology or community" (p.183).

A garden may seem small within the language of problems, solutions and outcomes of mainstream teacher education. But in a changing planet, a garden helps create 'opportunities' for creativity, responsibility and leadership. A garden gives learners a chance to examine the underlying assumptions that have long promoted a view of learning severed from environment. Including such ideals in teacher education challenges mainstream learning at the very roots and brings hope that environmental ideals will have central, daily importance.

The movement towards a community learning process took place spontaneously. While the garden started with the notion of a 'model' outdoor classroom for university students and schools in the community, soon the 'learning' in the garden was about learning the garden itself and the learning gardeners themselves. All became part of a 'learning laboratory'. As participants in the narrative, we started out thinking the garden was about the 'environment', in a way that did not really involve us. Early ideas about the garden (i.e., 'we will create a sustainable environment' and 'we will teach students about that environment') exemplify a deeply detached perspective between land, learners and community. Such an ingrained, top-down, 'context-free' perspective also shows how easy it would have been to go on teaching and training teachers as if the environment did not matter, or even exist.

It was not until we began working and learning the garden, and made a connection to the land, that the idea of learning through connection (and community) seemed possible. In time, the notion of developing a 'connection' between land and learners was a priority.

Over time, I have noticed how deeper biases, epistemologies and cultural understandings all come to light in the garden. A garden is a place for sharing worldviews, as a community, in the context of ecological justice. Moving beyond a 'problem-solution-outcome' model of teacher education, the garden invites a form of inclusivity where 'alternative' models of learning and diverse cultures can lead students towards greater possibility and freedom.

The garden continued, and dramatically transformed, through the ideal of inclusive community.

The Fire Pit

> Time-honored values of respect, reciprocity, and cooperation are conducive to adaptation, survival, and harmony. Native people honor the integrity of the universe as a whole living being—an interconnected system. Since it is alive, all things of the earth must be respected because they also have life. Native people have a reciprocal relationship with all things of the universe. (Kawagley & Barnhardt 1999, p.127)

The founding cohort decided on a 'rotating stewardship' model, where each new cohort of students would take up, or challenge, the ideas and visions of the previous cohort. The founding stewardship cohort believed that this would ensure that the garden was not owned by any one group, or individual, and that all ideas would be respected and remain open to assessment in the context of the changing needs of the natural environment. For example, with sudden drought or local water restrictions, a cohort would have to decide on the fate of the plantings of the previous cohort. The same idea could be applied to food and community; with an influx of homeless to the region, the garden could be used to raise funds and provide sustainable food. Thus, each cohort was guided by a collective consciousness towards the garden which also responded to needs of both the social and the ecological community.

A learning community learns through alternative models and from those who support a majority consensus model of decision making. The word 'community' often makes the rounds in education, usually described as a mutually beneficial network, where there is often individual (economic or professional) benefit. Such a model is little more than an expanded idea of an individual success model based on ingrained assumptions about learning and knowledge. In practice, 'democratic' inclusivity must benefit both the social and the biotic community and requires humility, sensitivity and risk. Community, inclusivity and democracy are meant to include certain tensions and, from the perspective of individual gain, certain losses.

One student teacher in the garden found his own way to learn (and teach), which went well beyond the usual expectations of mainstream teacher education. The student found meaning in the garden based on an indigenous learning model, specifically through the En'owkin process and traditional Okanagan teaching. The student found inspiration in the process of En'owkin, where

> a community is the living process that interacts with the vast and ancient body of intricately connected patterns, operating in perfect unison, called the land...This idea of community, as understood by my ancestors, encompassed a complex holistic view of interconnectedness that demands our responsibility to everything we are connected to. (Armstrong 2005, p.15)

In consultation with family elders and faculty on campus, the student teacher and a student in Fine Arts built a traditional Fire Pit next to the garden that included ideas around the Medicine Wheel[1] and the notion of learning as process. The students built the project as a research project on a site nearby, at the bottom of a steep slope beside the pond, just outside the (initial) boundaries of the garden. The Fire Pit, a traditional 'bowl' carved into the earth, was meant to last for generations as a symbol of community and learning linked to reflective land stewardship. The symbolic value of the Fire Pit emphasized the idea that the entire garden, as a whole environment, was a site of deeper transformation. The 'outsider' knowledge of the pit became part of the garden.

The student researched, held an oral interview with an elder (his grandfather) about the Fire Pit and the teachings of the Medicine Wheel. In the Medicine Wheel, fire symbolizes renewal and is traditionally used to cleanse the spirit, to heat the grandfathers (rocks) for the sweat lodge ceremony; the rocks are also symbols of the earth's core, one of the seven directions. Educator/artist Shannon Thunderbird (2008) describes: "Fire represents cleansing and renewal, for out of the ashes comes new growth, new thoughts, rebirth of ideas and new ways of being; the plant world regenerates itself in a healthy way from the ashes of the old. Fire acts as a Messenger and is a gift from the Great Mystery." The Medicine Wheel is a core ancient value and spiritual symbol in indigenous learning based on seven directions (North, East, South West: Life above the earth, life on the earth, life below the earth) (Thunderbird 2008).

A goal for the student project was thus to "open up new directions" and to "look outside established pedagogy to see what other disciplines...had to offer educational practice" (Marchand, M.J. 2007, p.1). As the student explains,

> I knew I wanted to express the En'owkin process in a physical way in the garden. It needed to be something that could be touched and felt, something that children could work with and learn by interacting...I was thinking of incorporating the En'owkin process into the garden experience by having four posts at key points in the garden with a sign on them to explain each of the four En'owkin food chiefs and to give the students a lesson in traditional indigenous Okanagan wisdom. (p.2)

Throughout the process, the student remained open to the ways the land was teaching him, even as he worked with the land. When the student began carving the grizzly totem pole, the first post in the En'owkin model of learning, he found himself working with a discarded, rotten piece of wood with uneven, pointed surfaces. The student soon realized that the shape and texture of the wood was 'speaking' to him, and 'it was not a grizzly but a salmon'.

First, for me, was the carving. I had originally planned to carve a bear figure as it is representative of youth and vision, however I could not find the inspiration in the wood and it was not coming along...I began to lose faith in the carving project until I began to realize that it needed to be a Salmon, representing action, that it began to take shape. (p.3)

The student worked on the project he described orally as 'non-traditional—as the whole idea of the project was to look outside pedagogy'. The student focused on the 'reflective part of the exercise' and 'applying the principles he was wrestling with' to his project. In other words, the student was not interested in applying Western principles of education, or even of fitting in to a preconceived idea of teaching, learning and research. The student's teaching, learning and research was in the process of the work *coming to be*. The bowl was filled with large rocks that extended eight feet down; the placement of the rocks symbolized the idea that "there is more to the earth, and human existence than the here and now. In time it will sink, be forgotten, but perhaps found again many, many years from now. The burying of the pit is symbolic" (Witzke 2008). Above the rocks, at the top of the pit, they planted an Oregon grape, a native berry bush, and ashes collected from the Okanagan Mountain Park fire of 2003.

The Fire Pit, the student's involvement in building the piece, and the collaboration with others around campus began to have a dramatic influence on the pedagogical direction of the garden, moving it further towards a symbolic gift which offered deep ecological transformation to the academic setting, to local schools and to the community.

The student told me that the Fire Pit was his contribution, a legacy, and a sign of hope that future learners in the garden could reflect upon. The student worked and thought hard, built the Medicine Wheel, then left its tangible and symbolic presence to future, still unknown, cohorts. Like previous cohorts, the work involved both practical work and symbol; this time, however, the work, the Fire Pit, was represented in an alternative model, which simultaneously operated both outside and within mainstream learning. The Fire Pit made tangible an ideal learning model that can be 'separate but equal'.[2]

The student (now a teacher) and the art student (now a working artist) later visited the garden as guest speakers for another graduate student cohort and described the symbol of the Fire Pit in the garden as an artistic experiment that also opened mainstream knowledge to new ways of learning. The former student spoke to the practicing teachers about 'listening' to the elements and how he learned to trust his own instinct in carving the salmon. This decision, like the existence of the Fire Pit itself, interrupted a set pattern and permitted spontaneous ideas to enter the learning process. The co-creator of the project discussed the symbols of the Fire Pit, the depth of the pit and the symbolic presence of women around a gathering place, at the bowl. The students discussed how the pit, in its reminder of depth beneath the surface, was a reminder of the 'earth' itself. The Oregon grape, which thrives without water for a year, was a reminder of sustainability and learning where the students 'made their mark' in teaching sustainability in a way that was also ecologically sensitive.

One of the symbolic values of including knowledge of difference and mainstream teacher education is to help students 'think outside the box' while inviting students to reflect on the deeper meaning of 'inclusivity' itself. The teacher told the group that the Fire Pit was meant to invite the class to consider new possibilities and to enhance their own learning. The Fire Pit was not a token symbol of 'inclusive' education, but an embodiment of a world-view that could actively inform the mainstream. The artist spoke to the group about remembering the earth, how all of our actions take place in connection with future and past. The Fire Pit is eight feet deep. Their work is a reminder of something basic that is often forgotten in talk around sustainability and ecology—namely, the earth itself.

The Fire Pit brought lessons to student teachers enrolled in mainstream teacher education about the true nature of inclusivity, about thinking and learning 'outside' the boundaries, with a sense of responsibility towards the land and one another. The Fire Pit stood as a positive, tangible symbol about 'what counts' as knowledge in Western academia.

> In Okanagan tradition there is always the inclusion of 'youth' in decision making, symbolized as 'newness' and 'creativity'. We always need to make room for newness because we need to be creative when we come up against something that we can't resolve or that we haven't faced before. Youth's responsibility is to apply their creative and artistic prowess to coming up with innovations, new approaches and new ways to look at things. (Armstrong 2005, p.15)

The focus on community and responsibility in decision making, in including an individual's unique qualities within the larger community, was also a powerful lesson for the teachers which helped them rethink the idea of 'democracy'. The Fire Pit made future students and teachers wonder: What is the legacy I will leave as a teacher? What lies beneath the surface of our own work that might be uncovered by future generations? What, as teachers, is the legacy we leave our students?

The Fire Pit was a way of responding to these questions in a way that placed indigenous worldviews, previously excluded knowledge, at the centre of that response. Kuokkanen (2007) discusses the problems faced by indigenous scholars within mainstream academia:

> What is it in indigenous epistemes that does not seem to fit dominant perceptions of academic knowledge and of the world generally? It became apparent to me that indigenous epistemes are allowed to exist in the university, but only in marginal spaces or within clearly defined parameters established by the dominant discourse, which is grounded in certain assumptions, values, conceptions of knowledge, and views of the world...in conversation with other indigenous students I learned that many of us found it difficult to truly express ourselves in the classroom except in indigenous studies courses. (p.xviii)

The Fire Pit provided a space that helped indigenous students express themselves in their own ways. The student who built the Fire Pit reminded teachers that they must include all voices in the learning process, that all must have a real chance to participate in a 'real democracy':

> Real democracy is not about power in numbers; it is about collaboration as an organizational system. Real democracy includes the right of the minority to a remedy, one that is unhampered by the tyranny of a complacent or aggressive majority. (Armstrong 2005, p.16)

'Real democracy' in the garden included risk taking and social justice, as students explored learning that embodied a process-oriented model. Such a new model, inclusive of alternatives, inclusive of possibility, inclusive of ignored or hidden knowledge, would play a central role in influencing future decisions for the space. In fact, from the time the Fire Pit was built, a transformation took place that opened the garden to new voices and new ways of conceptualizing the notions of learning in and *as* place, within community.

> While the human mind is naturally focused on survival, community-mind can be developed as a way to magnify the creativity of an individual mind and thus increase an individual's overall potential. (Armstrong 1999)

The Fire Pit's focus on native plantings, on community, on leaving a legacy and room to grow became a model for future learners in the garden. The new knowledge challenged teachers to include the idea of 'uncertainty' in their teaching, to simply leave room for the (as yet unknown) needs of future learners. The idea, the 'uncertainty', of providing a separate but equal space for new knowledge in teaching challenged a product-oriented model focused on the outward appearance of order and 'democracy'. The Fire Pit taught through its own presence (above ground) and absence (in the eight feet below); it taught teachers to be watchful for hidden lessons, for the less obvious ways we teach and learn.

A garden invites stewards to participate with heart and by giving voice to the land. This can be a difficult concept to engage with in the mainstream classroom, or even through text. The Fire Pit spoke large and with heart about leaving a legacy, and space, for future learners. The student teacher who built the Fire Pit was more concerned about what the space would teach, instead of what it would produce. In order to achieve this goal, the student had sought indigenous learning traditions outside the mould of mainstream, goal-oriented, performance-based teacher education. By doing so, he reintroduced the garden to process-based learning and community in 'real democracy' by expanding both the physical and pedagogical limits of the garden. In the many conversations we had together in the garden, we agreed that the garden was more than the soil, beds and plants. The 'earth' itself became the pathway of learning. The student also expressed his idea that the Learning Garden was really *a garden of the mind*.

A community model of learning was embodied at the indigenous Fire Pit as an extension of a belief system that existed both inside and outside of mainstream teacher education. Community, sharing, positive interdependence and deep respect for the ongoing benefits of agricultural land, instead of sole ownership or 'success', is part of that belief. The Fire Pit taught more about impermanence and teaching and learning as a 'gift'. While the garden was a rational response meant to address responsible land use as an alternative to golf courses, or real estate development in an agricultural valley, deeper notions of learning, ecology and community took time.

To develop an alternative model alongside the mainstream, even if that 'alternative' is based on ancient tradition, requires teachers and learners to move with agility between different worlds. Such movement itself is beneficial, as it is part of developing a creative alternative. In developing eco-centred education, it is not enough to work in ways that simply promote a prescription for sustainability; the students also needed lessons in community, caring, environmental awareness, mutual respect, problem solving and respect for alternative ways of working and thinking that did not always fit within their comfort zones. Through the Fire Pit, the student made it possible for 'what is not known', or for what is not normally accepted, to become part of a larger view of knowledge in the garden.

While I struggled to introduce various cohorts to the 'community garden' model, which was new to many student teachers in the region, the contribution of the Fire Pit provided a live symbol for a positive model of shared learning. The Fire Pit not only made tangible the idea that indigenous learning could be made central to the mainstream teaching of the garden, it set a new direction, in fact, it led the direction of learning in teacher education. The Fire Pit, the inclusion of outsider knowledge, advanced the collective vision by encouraging all learners to branch out and widen their ideas and projects on teaching and learning. The creation of the Fire Pit symbolized emancipation for all.

The student's salmon wood carving—and, even more importantly, his presence and the gesture of the Fire Pit as a gift to the garden—continues to help student teachers realize an elusive pedagogical principle: that the 'outsider', the non-mainstream presence, teaches by actively, and tangibly, challenging ingrained practices around 'insider knowledge'. The indigenous, 'alternative' garden, in its location beyond the boundaries of the garden, symbolized a wider range of learning for the garden that included local/global community, diversity, freedom of thought, ecological and social action and, for the wider education community, transformation.

Second Summer

I have noticed that when we include the perspective of the land and of human relationships in our decisions, people in the community change. Material things and all the worrying about matters such as money start to lose their power. When people realize that the community is there to sustain them, they have the most secure feeling in the world. The fear starts to leave, and they are imbued with hope. (Armstrong 2005, p.16)

That summer yet another stewardship of graduate students arrived in the garden. The new group, a small class of practicing teachers and student teachers (post-practicum), decided a simple theme was necessary for the garden: community.

Field trips to local community gardens included discussions around ecological issues: how can teachers help their students step out of the 'star' model of evaluation? How do we train students to listen to what nature teaches? How do we work around the power models of education that validate teacher control and invalidate difference? The students were careful to model an open style of discussion; we 'walked the walk'. As we gathered around the Fire Pit that summer, it seemed more important to be open to questions than to find answers. It was now more important to reflect than to continue building. After the busy time of the spring, the garden needed emptiness and clarity. Learning turned back to the learning process itself; we made sure that while we questioned, we created the conditions whereby answers might be possible.

The intention of the course, and the garden, was to make sure that teachers had a sense of community and the heart needed to learn from the land and from one another. From there, ecological and social justice would also have a chance to bloom. At the start of the second summer, there was a new, slow development towards a sustainable, and thoughtful, community of practice.

We began with *the gift*. The summer students stood in awe of the gift of the garden created by previous cohorts. They complemented the hard labour, the love and hope that must have been needed to develop the place. The group wanted to maintain the work of past stewards and reflected on its teaching. The intensive campus land development, the destruction of surrounding grasslands, made students realize that the garden was located in a rapidly shrinking island of pine forest. It was the 'summer of dust', and the group decided to spend the time writing, in community.

That summer, the students reflected on teaching, learning and possibility. In the garden, the graduate students (all working teachers) reflected on the land development, on the pine forest, on grasslands and on the role of themselves as youth, as teachers, land stewards and witnesses. The students took up the legacy of the Fire Pit and the garden and wrote poems about ecological justice, risk taking and transformation. These 'outsider' positions, in honestly listening to their surroundings, became part of the garden community.

Okanagan an oasis of mountain respite.
Layers of blue, gray horizons fade into hazy skies

To sleep and sweat on sun swept gold sands
To wash the heat away in a deep blue plunge
To remember a time when sun and wave guided each day.
(DS, From *Okanagan an Oasis of Mountain Respite*)

I am thinking of Gilgamesh,
When the World was Young.
The cell phone rings.
The beep of a text message.
The noise of a construction truck
building a new home for rich people.
And I am in the park
by the lake.
(YK, From *Gilgamesh, When the World Was Young*)

Put your juice-box straw wrapper in a little bag
Put the little bag in a bigger bag
Put the bags in a box
Put the box in a hole.
Problem solved.
(MB, From *Swearing and Litter*)

Learning in the garden that second summer brought new forms of ecological and community connection. Their clear-eyed poems opened the Learning Garden to even more possibility and chance.

Cascading down the mountain side,
set out in seemingly endless rows
a miniature forest of vines
cling to one another.
(LB, From *The Vineyard*)

As the students shared their own narratives and poems about the transformation of the garden, transformation took place in the students. In the second summer, a poetic awareness of the land allowed students to learn in an organic process without always knowing exactly where the lesson would lead. Learning in the garden meant rethinking notions of land and ownership while emphasizing the idea that all teaching and learning, all students and teachers, requires thought and care.

The second summer, the students and I watered and weeded, wrote poems of witness and were brought back to the beginning of the cycle: the garden as environment. We noticed how the Oregon grape, the native plant at the Fire Pit site, flourished, even during an intense August drought. Walking back and forth with the watering can, we realized the hardy little plant with its dark green leaves showed both a *practical* and *symbolic* way for future cohorts. The stewards decided that all new plants should focus on native species, as these mirrored most closely the values of the garden. Like the garden, the students' poems were small-scale, authentic, honest expressions of local goings-on. As the language describing the garden changed, in writing in vernacular voice, so did the notion of 'beauty'. Were the water-hogging roses really beautiful? The little Oregon grape stood, green and alive, on the dry forest floor.

A garden always brings learning back to the garden, to community.

Conclusion

How did we, as educators, as earthlings, get to the point where the environment is normally excluded from decision making? The reclaiming of both land *and* community is a relearning of both. There is potential in the experiment of a garden because it exists in the context it describes, in an actual physical place that directly influences how, what and where students learn.

A garden directly challenges the systemic myth of 'ownership' of knowledge since a garden is always rooted in shared knowledge, both as environment and as community. A garden could not exist without community, as it is impossible for one person to build and maintain a garden; it is also impossible for a garden to grow as a pedagogical process without the influence of ongoing stewardship. The Learning Garden teaches teachers that even the striving to develop community brings knowledge.

In time, the garden began to focus on sustaining a practical land-learner connection and on developing inclusive community. The garden, and teacher education, developed through a discovery of alternatives. Pedagogical understanding did not grow from 'limitless freedom' but through finding ethical ways of working, which included limits. The understanding of human limits involving the land defined our ideals around both environment and pedagogy. The garden taught the need for permission and the acceptance of limits. Working in a garden thus taught teachers to approach the land in the same way they might approach teaching, in terms of a land ethic. As Aldo Leopold (1949) notes in *Sand County Almanac*, "A land ethic of course cannot prevent the alteration, management, and use of these 'resources', but it does affirm their right to continued existence...in a natural state" (p.203).

The idea of maintaining a basic level of integrity, of a 'natural', 'organic' state of an organic compound, of not pushing the limits of nature past the point of renewal, also holds lessons for teachers. As Leopold says, "An ethic, ecologically, is a limitation on freedom of action in the struggle for existence. An ethic, philosophically, is a differentiation of social from anti-social conduct" (p.202).

In the 'struggle for existence' in the garden, in learning how to become a teacher, a garden has a grounding effect where students learn through both practical work and open community processes. A 'garden' model of teacher education teaches through 'real democracy' and process. In fact, being able to learn in combinations of intangible and tangible processes, individual and community knowledge, pedagogy and practice and the idea of freedom and limitations is part of the garden 'learning to learn' process.

As Orr (2004) and others have pointed out, education can no longer rely on the propagation of abstract principles taught and learned in isolation. I interpret the phrase "all education is environmental education" (p.12) to mean that learning begins with environment and community, whether those realities are acknowledged or ignored. Teachers lead by telling the truth, by acknowledging realities of diversity and the existence of alternative ways of teaching, learning and knowing. How else can teachers lead learning, in community, in reality, except by learning themselves?

Teaching and learning in community also means to dwell in an ongoing experiment where both teacher and learner are open to the ideas of others. Experiments do not always lead to profound changes; however, a stance of openness itself changes our encounters in teaching and learning. The idea of new teachers incorporating a state of vulnerability into their daily practice takes qualities of maturity, courage, wisdom, confidence and a profound sense of both individuality and community. The community itself must provide a space where individuals can be vulnerable and grow in these qualities.

Some former students/teachers have become an ongoing part of the Learning Garden and have involved themselves in the place after the course work ends. Former students return to talk to the garden stewards, sharing their interpretations. In their return, they take part in authentic learning community, a place where students visit and (voluntarily) connect. The garden truly inspires a space where 'all education' can become 'environmental education'.

When the students returned to talk about the Fire Pit, it was obvious that certain ideals in their work (process, connection, land, community) now embodied the garden's pedagogy. Their return was part of the process of including 'outsider' status; their work was no longer external, but physically and pedagogically embedded.

The move from building a garden to developing community is a process of inclusion, an embrace that moves outward. A garden facilitates learning principles that mirror physical, organic realities and thus invites a real possibility of moving away from hierarchical notions of teaching and learning that prevent even the possibility of authenticity and growth. A garden strongly supports students who are 'the exception' to the model of leader/follower that excludes 'outsider' knowledge and notions of difference in a predictable consumer society. Communities of 'difference', ways of knowing that have been marginalized in academia (including manual labour, collective decision making, local community, the notion of 'small-scale' solutions and learning encounters), find a place in the garden. In the garden, a community of difference is not just 'included' as a footnote but actually leads an outward movement to larger ideas around teaching and learning. By focusing on nature's economy we are making what was invisible visible. Working close to the land, we are aware of a community in need, of a larger education system that needs community and transformation.

A garden constantly extends outward: from garden to community to collective thought to changes in ideology and perception. The changing 'encounter' of a garden process provides profound lessons as the garden moves from 'outdoor classroom', where the goal is to 'get the job done', to 'transactional community', where goals are embedded in the process of listening to exceptions to 'the norm' that can positively influence and enliven the learning community as a whole. As groups of student teachers appear and disappear (and reappear) at the garden, transformational processes quietly unfold.

The metaphor of garden as community helps focus on local action, global impacts and the ways in which individual actions affect ecological and social communities. The garden as community supports Shiva's (2005) notion of "Earth Democracy", where "No humans have the right to own other species, other people or the knowledge of other cultures" (p.9). The garden extends

Shiva's principles and other writings around global peace and community by sharing land, seeds, community and ecological ethics of care; in doing so a garden helps promote the conditions by which sustainability can take place. In time, the measure of our community involved a transition from concepts of environment as physical landscape to concepts of community, the struggle for which lead to a new metaphor: transformation.

As one of the first garden students mentioned in the pond discussion, "nothing is really separate if you support the general idea of it." Such non-linear wisdom challenges the compartmentalization of knowledge. Even the transition towards community did not 'replace' the emphasis in the garden on the actual environment; instead, one idea informed the other in a non-linear, unpredictable, reciprocal process. Such 'non-linear' pedagogy quietly transformed the entire campus. Over time, no one could look at the pond and think of it as separate from the ecosystem around it.

The notion of teaching as a system of production diminishes the learning process. If learning were ever 'complete' in a garden, a garden could no longer exist. Despite all of our visions, ideals and theories, the garden, and our knowledge, grew in unpredictable ways. In teaching and learning, we need a garden's unpredictability, its social and ecological community, its life and possibility. A garden teaches how an authentic learning community, one that grows large through diverse knowledge and difference, is a catalyst that brings renewal in its own re-creation.

> A Medicine Wheel can best be described as a mirror within which everything about the human condition is reflected back to each of us. It requires courage to look into the mirror and really see what is being reflected back about an individual's life because some of it is painful...while [some of it is] joyful and reflective. However, it is ultimately facing the pain that makes each of us a stronger and better person. (Thunderbird 2008)

CHAPTER FIVE

❀

Garden as Transformation

Introduction

> I rejoice that there are owls...they represent the stark twilight and unsatisfied thoughts which all have. (Thoreau 1854, p.88)

The third metaphor for interpreting learning in the garden, transformation, focuses on physical transformation (from junk pile to organic garden) and on transformation of the student and the learning process.

Inspired by previous cohorts, and by the ideals and principles of the Fire Pit, by the fourth cohort a new metaphor emerged for the garden: transformation. Transformation, in the garden, arrived through shared narratives of ecological and social justice that involved letting go of individual success models and anthropocentric views. Letting go of such views was in fact part of the transformation.

The potential for transformation in the garden among student teachers is not simply a personal transformation of individuals, but a transformation that directly involves students learning to learn in organic process, within nature. The framework of a garden permits inclusivity of voices, as students speak of their transformation through the land. In a garden, the transformation from student to teacher potentially extends even further, as students learn to speak *for* the land. In my role as facilitator in our garden, I witnessed how, in turn, the land teaches learners how to speak.

How does a garden bring transformative awareness that leads to a shift in learning? How does transformation, in a garden, in community, shift the emphasis from teacher to learner to land, revealing the transformative process of learning itself? This chapter describes the potential role of transformation in learning in gardens. The chapter specifically focuses on lessons from the fourth cohort of students and includes excerpts from student writing to explore the transformative process and narratives of eco-centred teacher education.

The Possibility of Transformation: Setting Up the Practical Conditions

> Among the most daunting challenges of our era is the task of bringing about
> the transformation of consciousness that will be required if we are to move
> away from a culture predicated on consumption. (Smith 1999, p.207)

What happens when students become teachers in a garden? A fourth cohort of student teachers began in the fall. This time, I consciously set out to nurture the practical learning conditions by which the garden could invite community. I redesigned the course work to include cooperative, hands-on garden-based projects, where students team-designed curriculum. Students were also required to participate in community service and write Field Notes from both a service site as well as a nature site of their choice. The Field Notes would form a record of the students' 'transformation' from student to teacher and at the end of the course they would present their notes to the class. Some students chose the Learning Garden as their nature site, others chose a park, a hiking trail or the balconies of their apartments.

The team-based curriculum involved the development of individual lesson plans in student subject areas, plus an interdisciplinary series of lesson plans for the garden. The teams included a 'core' group of four, assigned to a garden box, who were further required to co-develop a portion of their plans with a team in an adjacent box. Inspired by the dramatic, outward movement of community in the previous cohort, I hoped to develop an expanded interdisciplinary, eco-centred, inclusive, community model of learning.

Once again, the work challenged their expectations about 'how to' become teachers. Some were surprised at the amount of group and interdisciplinary course work required in the course. Some resisted the perceived complexity of the work; the combination of community service sites, nature sites and the garden site was intended to invite students to broaden their notion of learning in community and simply help them define their own notions of 'a classroom'. Involvement in a kind of 'mobile classroom' invited students into an open-ended learning experience by situating their learning within local land and community. The challenges of developing a land and community model for teacher education mirrored the actual responsibility of classroom teaching.

During the fall cohort, one student told me that they would *rather* take soil or water samples and track quantitative data. The collection of scientific data in the garden was also welcome. More importantly, students would move from ingrained, binary ideas about learning (i.e., that students *either* take soil samples *or* engage in the processes of learning to teach) as they began to see how their practical knowledge obviously had a role in an interdisciplinary, integrated learning community in a garden. The challenge was to interpret the data in the context of a land ethic of ecological justice and in their own evolution as teachers/land stewards. Placing individual data collection in the context of community could also potentially 'transform' their own notions of learning that emphasized abstract learning and research in isolation. For example, what if the pond was full of contaminants? Would that information bring a responsibility to investigate the ground water and use of fertilizers/pesticides on campus? What was the responsibility of providing a space for migratory species, even as humans altered the biotic compounds of that space? The garden began teaching teachers that 'data' has application in the real world, from scientific, ecological and ethical perspectives.

The fourth cohort's assignments were, more so than in previous cohorts, focused on developing eco-centred curriculum and creating an inclusive community between land and learners. The scientific facts of climate change, the permanent alteration of the local agricultural valley for real estate development and the systemic exclusion of indigenous students from mainstream learning are vital reasons to develop teacher education focused on ecological and social justice. Labels of sustainable or eco-centred education without altering the settings and processes of learning simply reproduce the same processes that have for so long ignored the environment. Teacher educators in the midst of climate change, land development and the 'green' language of sustainability hold a responsibility to 'walk the walk' on environment.

Introducing a garden to student teachers requires initial discussion, and a rationale, on the way that teachers' work is embedded in community and in a (changing) social, economic and natural environment. For these reasons student training must include cultural and ecological awareness and hands-on explorations that will help them make informed decisions. Students in the garden worked in groups and were challenged to make decisions through consensus. Their participation marks for the course include a percentage graded by (an anonymous, randomly chosen) peer.

Many of the students in the fourth cohort reacted positively. The group included a large proportion of relatively younger students (average age 24) who were deeply concerned about the environment and were ready to take up the challenge. This group included students with experience in outdoor community learning such as 4H and camp leadership. Some had read about the facts of climate change, the limited amount of fossil fuels and food politics. Some were keenly interested in the environment out of a concern for what they saw as an increasing social and ecological detachment brought on by a booming real estate market and decline of small farms. The students arrived with varied academic interests and backgrounds, but in true community, they worked harmoniously towards collective action and awareness.

The Fourth Cohort: A Diverse Dynamic

A majority of students in the fourth cohort happened to have undergraduate degrees in the humanities. This group was vocal and critical about the shift in the valley away from small agriculture. Several of the students were from small towns around the province and had firsthand experience of a dwindling natural resource economy. They arrived from towns with sawmill closures and extinct coastal fisheries.

Ironically, while some students initially wrote about "not really thinking about or even really liking nature" the majority of students were excited about the social/spontaneous possibilities of the garden. The narratives of the student Field Notes became a personal record of the transformative process that occurred both in the garden and in the student teachers.

> Raising my ecological awareness should not be a problem...For me vegetables and flowers come from Safeway. However I am eager to learn. Amazingly I grew up in a farming community...yet I have very little experience in the garden.

There was also the hope provided by the Field Notes assignment itself:

> The Field Notes assignment gives me the opportunity to remind myself of the importance of empathy in developing my own understanding of social and environmental justice. It sort of formalizes my awareness of empathy as a basis for instilling ethics in my future students. Examples of social justice, environmental justice, empathy and transformation are found in my reflections on my nature and service site.

The assignments were designed to allow students to make their own connections in the garden; they were free to make a wide variety of interpretations that linked the work of becoming teachers with new awareness of the land. While the work might have seemed complex or highly abstract inside the classroom, as soon as we went outside to the garden, and the students realized they could make their own decisions, they began to see the garden as an exciting alternative. They were energized by the idea of directing their own teaching and learning journeys in connection with the land.

The idea of learning to teach by learning to learn, through the land, was a dramatic departure from most of their previous undergraduate training; such ideas were difficult to describe to the class before we went to the garden. Even the best lecture or textbook could not clarify a project about involvement that itself depends on a student's involvement in those very connections. In the garden, a student's willingness to engage in the environment, both as individuals and as community, was vital in developing eco-centred teacher education. When the students made their own connections, they were also beginning to take part in a learning process that challenged abstract, inflexible, non-participatory, subject-specific outcomes and rigid curricular goals.

Through their work and writing, students learned the paradox that ecologically and socially just teaching and learning is both a process and a goal. Inspired by the freedom of the garden, students questioned previous, decontextualized models of learning; they were enabled to view teaching as a 'movable space' that could place in the wild or in any setting that inspired ecological, ethical models of learning and community.

The best part of the process, from a teacher educator's standpoint, was that in the garden, student teachers changed how they learned; they could include local and global environmental concerns in their notions of teaching and could make ecologically literate design decisions in their lesson planning. The students created conditions that would inspire empowerment in their future students. The garden was not teaching them to teach in a garden, but in place, and in many places. It was powerful to watch the students learn to learn.

During their undergraduate training, the students had been given little opportunity to write about or engage in their own learning process. Writing the Field Notes thus gave them an opportunity to uncover their own biases. One student, a 'star' undergraduate in the humanities, wrote about how openness to challenge (in this case, a challenge to incorporate the land in learning) also brought a chance to develop deeper knowledge. The student connected that awareness to his larger goals as a teacher:

> I have to say that when I first read the outline for this course, the garden focus of the class and assignments worried me because I was not a gardener...I'm a guy who likes football and books about war; I don't garden or

write in journals. As I wrote the previous sentence I just realized that it is a very poor, close minded attitude. Wow, as a future teacher this is probably not the right attitude to have. It is easy to forget that my favourite classes in the past were the ones that were the most challenging...I suspect that I may run into the same attitudes, thoughts and feelings that I just displayed with some of my future middle school students. It will be important to impress upon them to be open minded and give the endeavour a chance.

Through the Field Notes, students began to shape their own challenges, in connecting with the land, in opening their minds and in reconceptualizing the notion of 'becoming a teacher'.

Native Plant Species: Criteria and Rationales

The students developed their first lesson plans around the previous steward's principles on native plants. The students' plantings/box design were considered in light of the original criteria set out in the first cohort (i.e., an all organic garden) with new ideas based on current research that potentially introduced

- more native, drought-resistant plants;

- plants that would help teach the difference between the water, soil and nutrient needs of imported plants versus local plants (i.e., the varied water requirements for English roses versus Oregon grape);

- plants that raised awareness of issues related to sustainability and small, independent farming traditions (i.e., cultivation of heirloom tomatoes, possibly selling heirloom tomatoes to local restaurants, collection of heirloom seeds);

- plants/designs that grew well with little water that could also be used to support the social community either through donations to the food bank (i.e., zucchini, squash) or donations to schools (i.e., pumpkins);

- more community-based activities that would raise campus awareness of the garden and pond (i.e., harvesting/donating produce for community causes, holding harvest sales on campus, building cob structures, developing whole class unit plans).

The students began by going out to the garden and examining the inspired work of previous cohorts. They spent time reviewing the lessons of the Fire Pit. They then turned back to the garden's often repeated ecological intelligence design motto, from Wendell Berry ("What is here? What will nature permit us to do here? What will nature help us to do here?"); later they redesigned the planter boxes to include more of a mixture of native, drought-resistant plants. Their rationale was based on local conditions: the risk of drought, the pressures on water due to real estate development, the rapid increase of new buildings on campus, the shrinking amount of available land for agriculture, the growing number of homeless people in the town who also needed fresh food, the lingering emphasis on imported (high water use) landscaping. Many of the plants they needed could be found at a local 'dry' nursery that specialized in native, drought-tolerant species and xeriscape. The boxes were thus redesigned so that they would link thematically through the plants themselves, though each box had its own unique theme

All of the student-designed projects and smaller activities held a purpose for the garden, for the community and for the learning process, as students documented their work in the Field Notes. As soon as the students took up the hands-on work in the garden, they began noticing sensory details at their 'nature sites'.

> It's Planting Day at the Learning Garden. It is busy up here, with noise bustling all around. There are several smells all over, particularly the smell of lavender being passed around from box-to-box. People are getting their boxes all set up...it's beginning to look great.

The various groups took different routes to the assignment; some strived for adherence to native plants while others focused more on developing a social network of trading plants with other groups. Some decided on themes of drought-tolerant food; some focused on developing aesthetics within the parameters of the assignment. One group decided on a theme of 'war on colonialism' that would divide their box and test the survival rate between local and imported plants. All of the groups' rationales were written with thought and care for the larger context of the learning assignment: the garden. The groups' conceptual diversity was made visible in the biodiversity of the boxes themselves. While students' goals for the boxes were rooted in farming/gardening terms, it was obvious that their 'botanical' dreams for the garden mirrored their educational goals.

Summary of Student Surveys

The students followed a mutually decided set of criteria for the box planting: drought-tolerant plants, native species (drought-tolerant, if possible) or plants that could be used to support community (i.e., for donation to the food bank or homeless shelter). Within the criteria, groups decided to name their boxes around themes that would promote certain ethical values and ideals they wanted to remember in their teaching. For example, "The Circle of Life" included the placement of a small, dying "Charlie Brown" tree one student brought from home; native plants were set in a circle around the dying tree as a symbol of mutual support.

After planting day, the 32 student teachers filled out surveys in their core groups of 4 and described how the decisions for their boxes were guided by a need for balance: in plants, in community, in teaching and learning.

Describe Your Group's Plant Choice

The surveys show group decisions made with consideration for community: how could the harvest be used to help the community? How could the native plants teach students and teachers? While students made choices within the limits of the assignment, they also began to understand the paradox of the 'freedom' that occurs within 'limits' and how practices that honour the limitations of nature support the continuation of life:

> Our plants are mostly drought tolerant with the exception of the strawberries. Depending on how many berries the plants produce we thought that we could either donate them to the food bank or share them with the classes who visit the garden. We also wanted plants that were aesthetic, so we chose flowering plants. We wanted to be surprised in the spring so we planted a row of plants where we were unsure of the variety. (The Garden of Educ)

> The purpose was for monetary and edible donations for the food bank. We can sell the lavender and donate the herbs and garlic. (Protector of Edible Beauty)

In their choices, most of the groups referred to sharing, to community and to future garden stewards. Through the planting assignment, students discovered the beauty of balance:

We divided our box in half. One half is composed of native plants. The al-
pine strawberries are ever-bearing and will provide food for future garden
stewards. The native flowers indicate the attractiveness of our native flora. In
the other half of our box we carefully transplanted a few existing plants.
These are mostly for beauty; but they also show the possibilities of transplant-
ing while recycling and honouring the work of stewards before us. (Survivor)

We wanted a natural, colourful, drought-tolerant butterfly and hummingbird
attractor. We wanted an aesthetically appealing garden, within the boundaries of the assignment—which was about the limits of nature. (Butterfly &
Hummingbird Garden)

One side produces for the food bank; the other side is drought tolerant, native species and includes herbs that can be used as medicine etc. (Work in
Progress)

The entire design of our box is focused on a sad looking little spruce tree.
We encircled the tree with drought-tolerant plants. (Circle of Life)

Students discovered the beauty in the local, and in the process of establishing,
not just a garden, but a process of place that could be revisited:

All of our plants are native, except the herbs which were already in the box.
We chose to keep these herbs because of their usefulness. Our main guide-
line was that the plants be perennial because of their 'share-ability'—every year
they can be divided and shared. We also wanted to challenge the idea of
what 'native' plants implies (i.e., 'barren' or 'boring'). We wanted to emphasize the botanical diversity of the Okanagan Desert. (Perennial Pyramid a.k.a.
the Circle of Death and Renewal)

Describe Some Highlights of Designing/Planting Your Box

For this question, a theme emerged where students described challenges and
surprises both in the discovery of nature and in the discovery of community.
As they learned to respect nature, they unexpectedly discovered how they
gained new knowledge, simply by sharing:

Trying to stack our rock formation was a challenge, but worth it as it brings
an aboriginal symbol of life and a totem to our box. We had fun trying to
plant, without killing, or touching, the earth worms. (The Garden of Educ)

Converting one of our group members who 'hated nature' into someone who loves nature! We worked well as a group. The other groups shared, so amazingly, in a cooperative spirit. (Survivor)

We learned a lot about native plants in our research; people learned that they knew more than they thought they did about gardening. (Work in Progress)

As they enjoyed the efficiency of pooling their knowledge, the students experienced the relaxed, thoughtful pace of non-hierarchical, garden-based learning. They discovered how the application of practical garden knowledge brings a sense of freedom at doing something 'right' for the environment:

It was casual collaborative peer interaction and it got us out of the classroom. A bonding experience. (Protector of Edible Beauty)

We removed a massive thistle weed, that felt good. Getting dirty while learning was great. (Circle of Life)

Getting rid of the drip-hose was liberating. Harvesting tomatoes and giving them to the food bank felt unexpectedly awesome. (Xeriscape)

One group described the simple joy of learning outside, while situating their work in both community and in symbols of nature that require fearlessness:

In preparation for planting we visited...the oldest domestic garden in the Okanagan. The garden contains many heritage plants. We also liked having enough to share with others. Fun, sunshine, water, dirt. It felt good to be outside learning. We like to symbolize the 'dark side' of middle school teacher education—we're not afraid of the cycle of death and renewal. (Perennial Pyramid a.k.a. the Circle of Death and Renewal)

Did Your Garden 'Plan' Unfold According to Plan?

After students submitted their plans I used grant funding to purchase plants from a local 'dry' nursery (at discount). The students were incredibly grateful, even surprised, when they received their requested plants; but if the plants were not available, they discovered the generosity in the Learning Garden community as students shared within and between their groups. The unplanned learning (the generosity, the ability to design in the context of nature, the shared work and ideals between cohorts they had never met) was a powerful part of their design and learning:

We did not expect to have any plants on planting day...we were surprised at how many plants we somehow ended up with, thanks to people's generosity. (The Garden of Educ)

The easy going nature of our group allowed for plan flexibility and it naturally came together in the end. We gained some herbs from other groups. We found an onion left over from last year's cohort. That felt like...a message of some kind. (Protector of Edible Beauty)

We did not have a prior plan other than bringing in wild native plants. The garden all came together on the spot. We wrote 'Do Not Water' with some rocks...and the symbolism of native plants vs. colonial plants, a lesson for future visitors to the garden (including us when we return post-practicum) all arrived spontaneously. (Survivor)

It went exactly according to plan in our plant choices. But the unplanned moments were the designing of the garden, which we created with what was already there. Otherwise, we started digging and let the garden guide us. (Butterfly and Hummingbird Garden)

Students learned through a combination of spontaneous initiative and organic design in the context of community support that included the value of gifts, 'mistakes' and critical questions:

Our box did not go according to plan. We didn't get our Evening Primrose, however people started giving us plants. So our box became a collaboration of donations; it was a social, community place with lots of visitors streaming to our box all day. Better than planting a primrose! (Work in Progress)

We couldn't find any of the plants we wanted. One student brought in the dying spruce, and everything fell into place after that. (Circle of Life)

Our plan was to just plant naturally and let the organic flow take over. All the plants we were given from others, from the instructor, and it was a huge surprise: we did not "plan" on generosity. (Perennial Pyramid)

Not being able to find local native plants at large local nursery was a surprise. We plan to investigate further. (Xeriscape)

Did Designing and Planting the Garden Change Any Preconceptions You May Have Had About Ecology-Based Teaching and Learning?

An overall theme of responses to this question revealed a previously held bias that a 'school garden' only concerns growing food, or aesthetics, and does not relate to issues in local land use or community. The students realized that a key part of the garden learning is socialization and developing a sense of interconnectedness that arises from the connections in nature itself. The hands-on work of the garden again provided a practical and philosophical framework where the students lead their own learning about the environment, community and transformative learning.

> The garden was transformed aesthetically, but the 'real garden meaning' also became clear. Group reflection is part of learning. (Protector of Edible Beauty)

The process of researching, designing and planting the garden as a community, based on a set of ecology related principles, allowed students to learn practical teaching skills and to believe in themselves as teachers and learners. Students realized they could apply previous knowledge and incorporate the learning of previous and future cohorts into their present work. They began to see the idea of a place as a process.

> We learned classroom management as the garden brings so much focus and interest. Students are not bored. Also, planting the garden brings about a sense of ethics and morality: we are giving to the community and not to ourselves. The variety was amazing: everyone had different ideas and stories about the garden. It was also cool to be involved in 'continuous learning'—we know we can always come back to this site, though we know it will change. That's cool. (Butterfly and Hummingbird Garden)

> We learned about xeriscaping that day. I was able to see the applicability of a garden to teaching in the classroom. It's about plants...but it's also about learning. The benefits of eco-centred learning were evident, especially in the sense of community we developed, and in the slow, but continual pace of our work. It felt like a normal, non-stressful way to learn. (Xeriscape)

Overall, the student teachers' responses suggest the importance of challenging, not just previous knowledge, but previous modes of learning. The students discovered how the garden, and the many alternate modes of 'learning' in a garden, permits a site that is conducive to deeper awareness.

> We were impressed with all of the other boxes and the ideas of our fellow student teachers. We were surprised at the amount of food you can grow in one box. (Work in Progress)

We noticed how suddenly everyone was involved—actively and individually—within their groups. We didn't think such an activity would be possible in teacher education, or that it would teach us so much about being teachers. We discovered how feelings changed and how the garden has healing qualities and helps solve problems. (Perennial Pyramid)

What Do You Hope to Teach Future Stewards and Learners Through Your Box?
Responses to this question revolved around teamwork and community which mirrored both a connected biological network and a thriving classroom diversity. In many ways, the notion of responsibility extended to both the social community and to the community of native plants.

We hope to teach future learners about the importance of teamwork as well as the importance of non-invasive, drought-tolerant plants. (The Garden of Educ)

We want people to think about teamwork; how the interrelation between all the different boxes is like an ecosystem working together; how transplanting/recycling plants and sharing resources enhances your own work. (Protector of Edible Beauty)

Students wanted future stewards to see uniqueness and a sense of imagination in the boxes, which would mirror their own potential and ability to learn. The students wanted to tell the future stewards to respond spontaneously and to let the organic process, instead of pre-decided plans, guide their actions:

Don't be afraid to see the beauty and toughness of your native environment! (Survivor)

Think outside the box...even within the limits of a box. Use your imagination. Collaboration and group work can result in something you may not have thought possible. (Butterfly & Hummingbird Garden)

We hope to teach future learners about xeriscaping and collaborative teamwork; this was about working with the environment and not against it. (Xeriscape)

The hope for the future stewards was that they would gain both practical knowledge and an idea of how the garden permitted an understanding of the complex interrelations present in nature.

> Responsibility, stewardship, environmental awareness, nutrition, group own-
> ership. We want people to nurture and respect living things. We want to tell
> people to never give up on the outcasts (i.e., the Spruce). Hopefully 'he' will
> thrive and hopefully be replanted somewhere else. If that happens, the native
> plants will have done their job. (Circle of Life)

Through hands-on engagement in ecological design, the stewards learned
lessons in how to interact and learn respectfully from within their environment. The project taught the student teachers valuable lessons in experiential,
"three-dimensional" teaching and learning:

> We put into practice the motto of our teacher education course with Dr.
> Gaylie: 'Three C's and E' (Communication, Creativity, Care and Empathy).
> We hope future stewards will learn how to recognize native plants, and learn
> what season they mature in. Future stewards should try to 'sculpt' a garden in
> three-dimensional design, not just two-dimensional. (Perennial Pyramid a.k.a.
> the Circle of Death and Renewal)

Transformational Process: Shift in Awareness

After planting day, students took a deeper interest in the garden. Some
students chose the garden as their Field Site, so they could monitor their work
in the boxes.

> Our garden has been transformed. We had a wild untamed jungle and now
> we have a conservative native plant paradise. It was a liberating moment
> when we removed the irrigation hoses from our garden. I never knew how
> beautiful a native plant garden could be and I must admit that my garden is
> beautiful.

Students came alive as they realized that their practice in class, their lesson
plans, their journals and, now, their campus were all a part of a web in the
learning process. As their knowledge of native plants and water use expanded,
students also expanded their idea of a learning process to include writing
about the new details in their surroundings:

> It is quite silent up at the Garden today, except for cars running on the re-
> cently opened road running along the side of the garden area. It's unfortu-
> nate that the road has been moved there. While it may be productive, and
> help the flow of traffic around campus, it is sad that it will ruin the tranquil-
> ity of the Learning Garden.

As students became more aware of their surroundings, those who had not been previously interested in nature, or social issues, suddenly began to wonder about the changes on campus. They recognized their learning space and took responsibility for stewardship; they looked beyond the garden and started questioning the larger development processes that surrounded the garden.

Student perspective shifted from small, practical details to larger issues:

> Already I am noticing what went unnoticed before. On my way out to the garden the first thing I noticed was the noise. As I walked from my class I weaved through groups of students and crossed a parking lot before reaching the garden path. I heard many different noises from the hustle and bustle of students to cars, busses and cell phones. As I made my way towards the garden, all of these noises faded and there was a brief moment of silence before a bird chirped and a cricket did his thing. It was a little bit eerie or odd, I guess. For a brief second there was silence and it was like I was crossing into a new dimension. Something was different. I also noticed that the air was fresher. As I think about my walk out to the garden I realize that I left a stuffy classroom, walked through two smoke pits and crossed a bus loading zone before reaching the fresh air of the garden. As I sat down to start to write and think about the surroundings, I noticed that the Learning Garden is a totally different setting; there are no disturbing noises and you can take a fresh breath. It appears life has slowed down. I can hear crickets. I can feel a slight breeze and the rustling of the leaves has a very calming effect.

After planting day, and in being able to write openly about their shift in awareness, students could see the links between the land and their new roles as teachers. The garden became an advanced lesson in learning to learn, through nature:

> I chose the Learning Garden as my nature site. I found this to be an appropriate place because of the class's involvement with redesigning and planting the garden box; I had an opportunity to watch how the boxes changed over a few weeks. Researching plants, using the guidelines for the Learning Garden, was a valuable exercise that brought awareness to the issues of sustainability and environmental responsibility. Using a garden to encourage students to slow down and appreciate the environment...would be a good place to start discussion on environmental degradation and desertification around the world, food security, stewardship and appreciation for diversity in their classrooms and lives.

For some, the shift from "I hate nature" to "nature steward" was profound:

Today, I experienced a strong transition in my mindset on the Learning Garden. When I first heard of the course work that we would be doing, I was quite upset. I didn't think that I would enjoy doing the gardening aspect whatsoever; as this is something I'd avoided my entire life. However, once we started, and I allowed myself to get into it, I was able to realize that it can be quite fun and relaxing. As the class went on, I got more and more into it, and eventually, I was able to come up with the strategy that my group ended up using for our box. I'm interested to see how the experiment works, with one half of our box being all native plants, and the other half being non-native plants.

After planting day, students more easily grasped the complexities of their assignments. The experience of the garden brought a deeper and more profound awareness of teaching and learning. It was the turning point in the class where students shifted from individualistic 'high achievers' to 'participants' in learning community. Even as they learned how to teach, they were also beginning to teach themselves how to learn. From that point forward, a new, organic interest in the beauty of the garden became a self-directed learning process that nurtured deeper inquiry:

There are also two big cucumbers, some red five-pointed flowers, some purple flowers, some orange flowers, some red flowers, some little blue flowers, some yellow daisy-like flowers and lots of little flies. As I write this list I realize that my knowledge of plant names is abysmal. My goal for the night is to research the plants growing in my garden and find out their names.

As awareness changed, the students' academic work habits changed:

Today I noticed that the garden is a peaceful sanctuary. It was surprisingly nice to leave the hustle and bustle of the campus for an hour and take in nature. I also walked around the garden and found a couple of peculiar plants. My favourite is something called a Fiery Barbara. I know nothing about it, I just like the name. I am feeling refreshed even though I was theoretically working for the last hour.

As students reflected, they examined their own previous attitudes about nature. The students, like their own future students, would be a catalyst for change in their own families and households:

Today's reflection is a little bit sad. I am learning a lot about water conservation through the planning of the xeriscape garden. I cannot help but feel a little ashamed of all the memories I have of my mother watering her garden.

After reading all of the dire warnings from Al Gore, David Suzuki and others, it is a little disappointing to think that all this time we could have had a beautiful garden with different plants with less water. I am, however, feeling motivated to educate my mother on native plant gardening.

Ecology and Service: The Love Connection

Today we need open-eyed gurus, helping humanity to reconnect with Nature, restoring the spiritual balance. (Rowe 2006, p.45)

A core component of the students' course work was to take part in some form of community service for 5-10 hours during the semester. Part of the task was to initiate community service which would help them stretch beyond their usual routines and ways of thinking. The students would eventually share this work in class presentations. Some students set up lavender and produce sales on campus and brought the cash proceeds to the food bank. The students created lavender bouquets; one of the students with an art background painted calendula flowers on seed packets.

Other students chose work in the community. One student decided to work with physically and mentally challenged youth, as she had little previous exposure to students with special needs. Another student volunteered to provide swimming lessons to young people with Down syndrome. Some volunteered at the local animal shelter, cleaning out the cages of lost and abandoned animals. Many students volunteered at the food bank, and became well-known to the local homeless agencies. A few students volunteered in seniors' homes. One student cleaned up garbage from the side of the highway. One student left the assignment to the very end of semester, and on a cold November day went downtown to read poetry with homeless people.

The assignment helped students become 'whole teachers', who could work as participants in community with an understanding of the linked dynamics of society. They created their own examples of positive, peaceful interaction, while seeing how small actions affect larger systems. The garden was one analogy for a larger ecological system. Now, their community service grew into another analogy, as they took part in maintaining a positive, socially responsible community.

I noted how the students approached their service site with the same humble awareness as they now approached the garden; they did not want to simply apply their own knowledge to 'solve problems' from the outside. Instead, they immersed themselves and became part of a larger and more diverse environment; they took part by first *listening* to the needs of community, from within community. Half-way into the term, through a combination of firsthand experience in the garden, of developing ecologically sensitive design, of reflecting on their own learning in their Field Notes, the student teachers applied a spirit of ecological justice and social justice to learning. Participation in the nature sites, the service sites and the garden allowed them to see how smaller actions, approached in a spirit of generosity, and love, positively contribute to larger ecological and social networks.

> My service site experience has put me in touch with a student who has a different background than my own. I have learned that you have to look at the student, put yourself in their shoes, and really try to understand how they feel. This exercise of empathy is difficult both intellectually and emotionally. I think the best thing to do is listen. Listening gives you insight into their life and helps you see the world from their perspective.

It was a hopeful, empowering experience to involve new teachers in finding a place for themselves (and their students) within the classroom and larger systems of learning. Through conscious participation and mindful reflection, students designed their own learning even as they transformed into teachers. While some of the students were at first reluctant to participate, once they actually started taking part, and reflected on their work, an awareness of deeper systems in teaching and learning took place. The students had grown from the solitary goal of *seeking knowledge* to involving themselves in a cyclical process of practice and reflection within changing contexts.

The students were learning how to teach and learn as adaptation, while nurturing ecologically and socially just systems around their work. Through learning how to become both teachers and land stewards, they were learning, from the very start of their teacher training, how to act in their environment in ethical ways, mindful of their social and natural surroundings.

As the students presented their work, they broke through their own previous conditioning about learning and social change. I was continually surprised at the many ways they chose to embody empathy. The change in the students' own awareness was profound as the 'Field Notes' helped students make the link between action, reflection, practice and experience.

In their Field Notes, students went on to make metaphorical comparisons between the variety of plants in the garden and the variety of future students they would encounter: the bright and showy ones that thrived and required constant maintenance; the ones in the corner, hardly there the entire year, but suddenly blossoming when you least expected it; the low-maintenance students who worked hard and survived. The ones that disappeared before the end of term. The ones where you could never remember their names. The list went on. Metaphors multiplied. The biodiversity in the garden led the teachers to seek out the diversity they would encounter in the classroom.

Through community service the students discovered something about the gift of learning and about the joy and empowerment of having something to give. It was a lesson I knew they would pass on to their own students.

The students wrote about their discoveries in their journals:

> I think going to a place such as this was a really beneficial experience for me, as I have always had a tough time relating to seniors...this experience has reminded me that it is important to teach students that a small contribution is still a valuable contribution; not everything needs to be spectacular, nor does every action require acknowledgement. Instead, I hope to use a community service project to demonstrate humility and responsibility to the social and environmental sustainability of one's community and world; here students can experience 'thinking global and acting local'.

The theme of connecting with past stewards arose through their community service, and they realized that future stewards would also be a part of their current work:

> The food was collected and donated to the food bank while the seeds and herbs were sold at lunch and all the proceeds went to the food bank. While we did not have much to do with the original planting of this harvest, it feels good that the food was harvested and taken to the food bank. I think the original caretakers of the box would have been proud that their produce went to a good cause.

One student described their growing empathy. The student became an in-class tutor, and eventually a 'big brother' to a struggling student. He discovered how a teacher can be a helping presence, simply through listening:

> My service site experience has taught me more than I ever expected. First, I have been introduced to an aspect of schooling I was always uncomfortable with. I was not ready to deal with this aspect of being a teacher. This is the part where you realize that the child's home life is chaos and in fact his school life is actually a place of safety. Unfortunately, in some cases school

may be the safest environment for some people. I now see why it is important to have an adult advocate/mentor for every student because sometimes there's none. It is difficult to comprehend that as an educator you are concerned with content, curriculum and learning but in reality these concerns are often secondary to the social issues.

My service site experience has put me in touch with a student who has a different background from my own. I have learned that you have to look at the student, put yourself in their shoes, and really try to understand how they feel. This exercise of empathy is difficult both intellectually and emotionally. I think the best thing to do is listen. Listening gives you insight into their life and helps you see the world from their perspective.

The final presentations for the Field Notes described profound transformation in the students. Their shared reflections located their individual work in the context of a classroom community. During the presentations, students described how they had arrived at the path to transformative learning. As the students listened intently to one another, often staying well past the assigned end time for the class, I realized that one of the reasons why the assignment had such an impact was that the students themselves designed it. Simply being able to discover new paths to learning in the world in caring and ethical ways created a sense of limitlessness. Some described how such freedom made them feel 'young and alive' as they remembered aspects of learning they truly loved but had, over the course of their undergraduate academic training, forgotten.

Listening and responding allows for personal connections to be made with your peers' experiences...this could give an adolescent the nudge that they may need to be able to think about something from another perspective...new teachers and seasoned teachers alike are presented with opportunities where they must step out of their comfort zones.

Thoughts on Student Teacher Transformation

The profound change in the students occurred in response to their engagement with ecology and community. How did learning through environment and community inspire the students towards transformation?

The Learning Garden stewardship committee began with a small group of graduate students, all full-time teachers, and student teachers in an environmental education class. By the fourth cohort, the garden gradually transformed the students' approach to teaching and learning. What is perhaps most profound is how students, with permission to engage in eco-centred, community-based learning, began to authentically transform their own attitudes, from within. While students put practical, scientific knowledge of the earth to gardening, they saw how that knowledge works, and changes, through organic and community process.

The process of students learning in a garden revealed the role of process, itself, in learning. The students had arrived with preconceptions about eco-centred teacher education based on a prior belief that learning in 'environment' required specific scientific knowledge. Some thought of teacher training as a subject discipline that involves learning how to write lesson plans, keep organized grade books and manage classrooms.

The teachings of the Fire Pit, and my own 'teacher-teacher training' from the previous cohorts, helped me to prepare students for the deeper complexities of ecology-focused teacher education. Since the 'subject' of teacher education is often 'the process', I already knew the students would have to become accustomed to vastly different processes in order to learn how to embed ecology and community in their lessons and teaching. As other researchers note (Kaza 1999), part of the teachers' challenge in 'environmental immersion' is allowing students to find their own ways.

In discussions on ecology and school gardens with a new group of learners, it is vital to describe both the 'big' facts about climate change and the potential for 'small' responses. Environmental educators on college campuses note: "The facts can actually turn people away from meaningful engagement with the issues. They say the situation is too big, too all-encompassing, too demanding, too scary, too big to tackle. If I don't address these feelings with my classes, I find myself looking out on a room of blank faces" (Kaza 1999, p.145). In teaching a course that combines social justice/ecology and activism, Kaza discusses how the students learn to support one another in group consciousness raising. She also notes her initial fears in introducing new forms of learning on campus: "I am nervous every time I begin another one of these explorations...what will come up this time? Will I have the courage to let the students struggle into their own insights? Will I have the strength to engage my own conditioned thinking as part of our process together?" (p.158).

In designing eco-based projects, students also investigated various tools that would help them communicate their ideas. I set up a 'web wiki' where students shared links and resources and where they could also 'speak' to the land stewards of past cohorts. My role was also to invite them to begin, and once they were involved I provided a supportive setting for them to reflect and share. There was a time for action, and, as cliché as it sounds, a time to reflect. Orr (2005c) comments: "we know that the things most deeply embedded in us are formed by the combination of experience and doing with the practice of reflection and articulation" (p.99). The process repeatedly brought deeper awareness as students confronted their own barriers to learning and realized that breaking down barriers, and moving beyond their hesitancy, was also part of learning. The students later described in their journals how they learned to trust themselves to learn, by learning how to respectfully interact with their environment.

Immersing students in ecology in teacher education is such a new approach, one quite different from my own learning and teacher training experiences. As a teacher educator, I also learned a powerful lesson in allowing students to become teachers through learning.

> As this is the last entry, I will again address how much my opinion on the Learning Garden, and on nature in general, have changed since the beginning of the class. I am very happy that we were given this assignment, and hope that it is something that stays in the Middle School programme for years to come.

> Social justice, environmental justice and empathy are concepts that in order to be taught must also be practiced. As teachers we have an opportunity to expose our students to concepts and realities that exist in our communities, country and world...as student teachers we have the opportunity to be influential in the development of another person.

Students learned about the long-lasting, ongoing effects of small-scale action and how lessons in the garden could apply to their future work as teachers:

> A garden needs the appropriate components to flourish including a match between the plant and its environment. Similarly, a student will thrive in the right environment. Although any one teacher may only have the student for a year...they are still part of the small steps that work towards the end goal. I was able to be involved in planting my group's box and I had the opportunity to watch some of the small changes that take place at the end of the growing season, but I will not be able to watch these changes all year, or even next year. Above all, the garden is a reminder that as teachers there will be times when we need to acknowledge the environment that a student, group of students or the entire class is experiencing before we can focus on learning.

Anyways, back to my transformation. I believe that I am more appreciative of gardening in general. I have grown quite fond of our little garden box and when I first enter the garden I find myself excited and running over to my box to check how my plants are doing...My favourite plant, the Fiery Barbara, which I chose because of its name, has undergone a transformation. This plant was just a green thing but now has these amazing red flowers. It is beautiful.

Some students were able to bring deeper understandings to their previous knowledge of sustainability by being involved in a community 'cause'; their new awareness also revealed to them their previous attitudes.

First, I think I have improved my ecological awareness. The main area I improved on was an appreciation of nature and a realization of our (human) position in nature. Prior to the reflections, I was fully immersed in the global warming lingo. I was aware of notions and used terms such as EF, the 'one tonne challenge' and the genius 'environment saving light bulb'. However, there was a disconnect between understanding and semi-believing in a cause or message and actually experiencing nature and full-heartedly believing in a cause. During my reflections at my nature site I experienced nature. This may seem weird but really I had abandoned nature.

By the end of the semester, students describe how they internalized the transformation they had nurtured in the garden. They no longer drove blindly past the agricultural landscape that sustained them in order get to their destinations.

Today is a sad day. As I sit down to reflect in my journal, I cannot help but notice the development occurring all around the Learning Garden. I was taking a picture and a plane flew overhead. As I glance back toward the parking lot I notice a pump truck doing something with the drainage system. If I look beyond it, I can see bulldozers moving the earth at the new dormitories. If I look to the right I can see residential housing being built on the hillside. If I look to the left there is paving going on at the new roundabout. And if I turn around I am looking directly at a house. I am definitely feeling squeezed in.

Today I am recognizing our place in nature. By our place in nature I mean human beings' place in nature. I know that we are only a part of nature but with the development that is going on around us right now I cannot help but think of humans as the ultimate invasive species.

I wonder how it came to be that nature seems to take a backseat to the idea of development, progress and growth. I am finding myself torn...I guess the irony is that in this idea or quest for progress we could ultimately be killing ourselves. I am also struck by the irony that a few weeks ago I had to go to the Science Centre to see an exhibit on native plants.

When students looked back on ways of working and being in environment that were wasteful or destructive, or caused permanent alteration, they saw themselves, and previous ways of knowing which had been shaped externally, through notions of efficiency, instead of within the wisdom of nature itself. The students realized there was 'something wrong' with the idea of unlimited growth and that their smaller interactions with the garden as environment, as community and as transformation also counted.

A comparison of Field Notes entries at the beginning and end of the semester shows how the students internalized the transformative process. Their earlier entries exemplify a sense of detachment from the garden and learning:

I am looking forward to having the nature site on campus because with my hectic schedule it will be nice to go to the garden between classes or after classes to unwind and collect my thoughts. It is weird to write that last sentence because I would think that as a lifelong student I would have done something like this before.

By the end of the semester, students explained dramatic changes in their awareness of how the landscape, and their learning processes, had shifted from a stance of observation, towards involvement:

All this reflecting on the idea of transformation has led me to conclude that transformation is an ongoing process. I had the preconceived notion that I would be a different person at the end of this exercise. However, I am not sure that there is a definite end point. I realize that I am *transforming*. In the back of my mind I am subconsciously aware that I am different and I am excited because it is for the   better. I am no longer focused on an end point.

Field Notes entries towards the end of the course movingly describe these internal/external changes:

I stopped my car in the middle of the roundabout at the entrance to the university. I stopped because in the midst of all the crazy construction that is taking place, as my once small college transforms into a much bigger university, someone took the time to plant a little garden in the middle of the roundabout at the main entrance...It got me thinking about how the natural landscape which once effortlessly existed has been torn up to make way for growth of the university. Despite the destruction, someone made the effort to restore some natural beauty in the middle of the constructed environment. But, looking even closer at the garden they planted, I thought, 'wow, how ordered and proper it looks'. Why are we constantly trying to re-create or redesign things in an ordered fashion? Even in nature, we attempt to make things appear as though they conform to our standards and expectations. Then I thought about how every time I go to the Learning Garden, I am not that attracted to the garden boxes, but rather I am drawn to the pond and the embankment where the pumpkins are growing so freely.

I like the pond because it exists, whether I give it attention or not. The pond is itself, whether I construct it, approve of it, or tell it it's beautiful. No matter what I do, the pond exists in its own beauty, and that is a powerful message for me. It makes me think about teaching and how I need to accept students as they come to me. I need to recognize that I cannot construct them, or shape them, or mould them to my expectations, but instead I need to appreciate them because they simply exist as they are. My students will come to me in a disorganized and unordered fashion, just like nature usually appears to be, but just like nature, they are all beautiful in their own way. It is not my duty to place them in a roundabout, organized in a way that makes sense to me and creates a nice perfect picture to the outside viewer. No, in fact, it is my duty to nurture and appreciate and respect them as they come to me. Disordered, natural and unique is how I want to see my students.

The Field Notes describe student understandings of controls on nature and controls on learning and how a garden, like a classroom, may only "appear as though they conform to our standards and expectations." By cultivating the garden, by getting close to the wildness that surrounded them, the students began to accept their role as teachers as a part of a large and diverse living system, even as they learned within the structures of academia:

I really like my garden box and it pains me to see all of the development crowding in on my garden. Especially now that the leaves have all fallen, you can really see the development through the thin sticks. I take solace in the fact that I now realize my own hypocritical ways and vow to improve them. I also have a plan for a native plant garden for my future home and if I get the chance I plan on joining or implementing a native plant garden at my future school.

The students truly 'greened' the university through their own experience of transformation in eco-based learning. As teachers they learned how their presence could cause change, for better, or for worse.

> When I first started writing at my nature site I was not sure why I was doing it. It is definitely weird in that I knew what nature was, I knew it was beautiful and had a shallow appreciation for it but I never took the time to realize the little things. I now realize and appreciate how peaceful it is. I have definitely transformed when it comes to my nature site. When I first started writing, my mindset was not appreciative of the assignment. Lately in my schooling I have kind of become caught up in the work. I appear to be in a rush to finish my schooling and join the workforce. In this rush I have kind of abandoned nature.

The garden teaches teachers how to teach and also how to be leaders who learn. Students quickly learn about nature's strengths and vulnerabilities; the idea of learning, and leading, through vulnerability is a new concept to them. In being vulnerable, teachers model a condition for learning in a spirit of openness and spontaneity. Leaders who lead through vulnerability create a condition for learning where their students are open to take risks; in such conditions, students become mindful and mature. Students learn how vulnerability also involves risks and leads towards transformation.

> If my students are the plants, the trees and the landscape in the garden, then I want to be the sun. I want to be the teacher who students want to be around simply because I bring them warmth, comfort, strength and growth. I want to be the power of the rays which nurture my students and make them feel better about themselves and their lives. I want to be the kind of teacher who students are effortlessly and naturally drawn to because simply by being in my presence, they are inspired.

In the garden, the students opened to new forms of learning and began to respect the concept of 'emptiness'.

> On the way up the path to the garden I stopped for a moment to take in the change. Every year these trees sprout leaves, bloom and then shed them. It shows how beautiful they can be. When they shed their leaves they bare their skeleton. You can see their soul. I picture myself getting too caught up in life sometimes. I become too concerned with material possessions...I kind of lose sight of who I am at the core. I think it would be nice to shed my leaves yearly and take stock of who I really am at the core. Then when I grow again,

I can grow better than before. I think I am going to...be like the trees and lose my leaves.

A garden teaches teachers that there will never be a point at which a student or teacher 'knows' all. A teacher's role is to open students to processes which will allow them to learn, and relearn. Again we learned the value of respecting all that we do not know. Students began to see the value in teaching and learning that is non-linear, eco-centred, process-based and, by necessity, always incomplete.

Transformation in a garden, as in a classroom, often occurs through unexpected and unplanned learning. The student's 'garden of the mind' did not involve a sense of individual ownership, but was a site of local/global community that provided the conditions where transformation could take place.

> Looking after a garden lends itself to a classroom analogy. You tend to the garden for a year and then you move on and the kids are in someone else's hands. I think eco-centered learning lends itself to teacher education quite nicely. Not only are we learning about ecological justice but you can really make the connection between care and respect. It is quite easy to read about care, respect and nurturing in textbooks but actually caring, respecting and nurturing a garden provides the opportunity to demonstrate, experience and improve these skills.

Stumbling Towards Transformation

New teachers often think that teaching 'well' involves having the 'right' answers; meanwhile, a campus garden promotes unscientific awareness of the natural world that includes the involvement of learners learning, imperfectly. In a garden, a humble disposition of *not knowing* is restored and learning begins. The lexical origin of humility is *humus*, a Latin word that denotes the organic component of soil that decomposes and breaks down. In the eighteenth century, with the rise of mechanized industry, the term somehow came to mean *lowly*. British Columbia naturalist and educator Stan Rowe (2006) describes the role of nature as an equalizer: "We are Earthlings, Earth internalized, humans from humus. Without the surrounding world of Nature our bodies could not be, nor our minds, emotions, languages, and cultures" (p.35). Learning to teach in a garden exposes students to the larger perspectives they need in order to teach.

Humility (empty, open, rooted in the earth) remains the entry point to ecoliterate knowledge. Ecologists note how such a stance of care, not control, towards the planet is the precursor to our survival as a species: "I believe that the evolution of a humane, just and durable civilization will require a great deal of love and competence and restraint that love requires for particular places" (Orr 2005c, p.103).

From the scepticism of students early in the term, through the native plant assignment, to student-initiated community service, to the transformation of their final presentations, the students thrived in conditions that permitted transformation. Turning to the land, students learned to act intentionally, and to learn and transform spontaneously. They learned to 'set the stage' of learning, then to trust organic processes and to take note. In the end, teacher education practice, combined with simple reflective processes, in the context of environment, allowed students to teach through larger, inclusive perspectives and humble awareness.

The challenges of one cohort triggered the transformation of the next. The actual site, the living place, passed from steward to steward while each cohort helped create, in the next cohort, the conditions by which students would learn how to design learning through nature. While students kept most of the plants from the previous cohort, they also learned the value of modifying past expectations in the reality of present needs. Students learn to waste nothing. Later, benches were constructed from fallen logs. Canopies were built from donated scrap material instead of buying new plastic versions from Wal-Mart. Like teaching and learning itself, students learned how renewal made sense in the context of a larger ecosystem.

Students began to approach teaching and learning with openness, as newly enlightened land workers, similar to the detailed observation in a farmer's awareness:

> I believe that whatever successes I have had as a farmer has come when I have approached my farm with what Zen Buddhists call a beginner's mind: without presuppositions, open to seeing and learning from whatever I encounter. Nobody ever told me about this way of learning when I started farming. Now I require apprentices to take a notebook and walk the farm several times a week, simply recording what they see. I want them to develop what I consider to be the most important agricultural skill—observation—and I want them to discover for themselves that biological systems never stay the same. (Ableman 2005, p.177)

Many students began their Field Notes talking about the actual garden as environment; gradually that awareness included community and ultimately a sense of personal transformation:

> When I think about my transformation, I know I am becoming more environmentally aware and ecologically sensitive. I am more in tune with concepts like my *ecological footprint*. I do not know how many times I have explained Xeriscaping to people; I think I am becoming an expert. Not only am I becoming more informed on the topic but also more appreciative of nature in general. I know it is such a cliché phrase but getting back to nature is something that I feel I am doing and accomplishing. I mean we are constantly being bombarded with global warming paranoia and I find myself in the supermarket trying to decide on the most environmentally friendly light bulb. But unless you take the time to enjoy and appreciate nature, I think there will always be a disconnect. I feel so far that my experience in the Learning Garden has helped me recognize this disconnect in my endeavors and I believe that I am working on eliminating it.

For student teachers, transformative awareness always means rethinking preconceptions of teaching and learning. Whereas the students typically arrived in the course wanting *to know*, and hesitant *to learn*, they gradually accepted a perspective on teaching and learning where a teacher never knows a specific end point for a lesson, but can only facilitate the conditions by which transformative learning might take place. Another new teacher awareness meant shared involvement over notions of individual ownership.

By the end of the fourth cohort, the learning community of the class had extended well beyond the actual class. It was an entirely new process as both a teacher and a teacher educator. By the end of the course students admitted they did not, and could not, 'know everything' about becoming teachers; nonetheless, a profound, difficult process of transformation had taken place. After the initial shock of environmental immersion, the students became extremely involved in developing new learning that directly involved the garden. The last day of class was a celebration of those learning paths, not as end points, but as a celebration of a transformative process that would continue into their schools. The final presentations on the last day of class went well over time, as students listened and stayed with each other. It now seemed more important to share in community, to nurture environment, than to brand themselves as 'good' teachers; the students realized that nurturing their environment permitted a new a path of teaching and learning that would simply lead to sustaining even more 'crops' in teaching and learning.

Like other teachers in other gardens, I was in awe at the students' ability to take up their own learning with so much thought and care for the land and for each other. They had entered the garden and embraced, even respected, all that they did not know. Approaching a classroom as an environment this way, filled with unknown learners, many of whom bring notions of alternative learning and difference, allows teachers and learners to teach in much the same way as organic farmers work with the land: respectful of spontaneous processes that support and sustain all surrounding life forms.

The first lesson of teaching in a garden is the way it encourages learning through process, in a constant interchange between concrete and abstract knowledge. As the students and I realized, in a garden, one minute you shovel compost, and the next you are struck by an intense, profound awareness of the interconnectedness of the life that surrounds you, the way the migratory blackbirds and other forms of life rely on the pond for food and shelter and how, in turn, the pond thrives as an intricate marshland because it sustains that life.

Learning in a garden, the act of sustaining the environment, soon begins to guide every action. And then, you shovel compost.

Learning in a garden relies on a blend of work, experience and reflection; it is difficult to describe the subtle impact of the actual interchange, of moving from hands-on work to a sudden, deeper appreciation for the subtleness of life and the many forms of diversity. However, in the garden I learned that it is not possible, nor possibly necessary, to pinpoint where, when and how an actual 'shift' towards transformation occurred.

Ecology-based learning simultaneously teaches process, community, knowledge and awareness. In this way a garden is both provider and facilitator of knowledge; a garden is knowledge and a garden is itself the embodiment of an ever-changing process of *a sharing* of knowledge, between past, current and future teachers and learners. In practice, the question *What does a garden teach?* soon becomes *How does a garden teach?* The process of transformation in a garden teaches students to be ready to learn from nature, with hands, head and heart.

Yarrow—Paprika
Chives
Gaillardia Goblin—Floramedia
Russian Sage—Salvia

Lavender
Rudebeckia—Brown Eyed Susan
Prickly Pear Cacti—Opuntia Humifusa
Hens and Chicks Cacti—Sempervivum
Echinacea—Purpurea Magnus
Calendula
Thyme—Thymus "wooly"

(From student Field Notes, list of new plants for xeriscape garden)

CHAPTER SIX

❀

Practical Matters

Introduction

This chapter includes activities, resources and other 'practical matters' related to teaching and learning in school gardens. The activities here are intended as starting points for educators and others who would like to incorporate ecologically and socially just curriculum and concepts in their teaching.

What are the practical conditions that will allow a school garden to grow? A 2006 Canadian survey of over 50 representatives from 19 secondary and primary schools and 22 school boards from across Canada found

- Most gardens are at elementary schools and usually have a garden committee with six members (teachers & parents)

- 90% of garden schools use the produce as part of teaching activities, class parties or snacks

- 70% of teachers teach in gardens twice a week or more

- 65% of gardens cost under $2,000 to start; 24% cost under $500 (largest costs were fences required by school boards)

- 91% feel school food gardens foster awareness of nutritious food

- Schools grew an average of five kinds of vegetables

- The most successful vegetables were potatoes, carrots, beans, tomatoes, beets, sunflowers, herbs, lettuce and squash. (Evergreen Foundation 2006, p.1)

A 2003 study (Kail 2006) found that, in Canada, 41% of 1,000 outdoor projects are abandoned. Some of the main reasons listed in the study are "difficulty maintaining site, inability or uncertainty among teachers about how to use outdoor classroom in their lessons, inadequate funding, vandalism...and loss of the site due to school expansion or relocation" (p.40). Research practitioners thus recommend that gardens should be easily integrated into existing curriculum, and that gardens be used as a central part of everyday classroom learning. The same report recommends that the planning, design and monitoring of the garden be left to students during class time and that "Maintenance must be considered before construction begins. If the need for regular watering during summer will present a challenge, consider container gardens" (p.40).

Researchers point out that a key area identified in the success of school gardens is the integration of gardens into curriculum. In turn, a successful curriculum also depends on how well theory is put into practice. A great deal of these decisions rests on the teacher in deciding on the needs of their students. However, there are models of GBL curriculum that can inform teacher practice, and a wide range of resources that teachers can use in developing curriculum that is useful, critical and eco-centred (Green 2004).

UN researchers found that successful school gardens around the planet share many of the same goals and benefits. In reviewing their extensive findings, researchers conclude that a garden pedagogy should include a wide range of skills development, which includes academic skills (i.e., "enriching core curriculum"); personal development (i.e., "add a sense of excitement, adventure, emotional impact and aesthetic appreciation to learning"; "To teach the art and science of cooking with fresh products from the garden or local farms"); social and moral development (i.e., "To teach sustainable development; To teach ecological literacy and/or environmental education; To teach the joy and dignity of work") (Desmond, Grieshop & Subramaniam 2004 p.49).

In the campus garden we constantly returned to ecological design principles which relied on non-invasive projects that preserved the integrity of the land. We returned to Orr's (2004) description of ecological design principles (i.e., "right scale; simplicity; efficient use of resources; a close fit between means and ends; durability; resilience" (p.105).

How did we put such ecological design principles into practice in our garden?

Three examples of ecoliterate design in our garden include a meadow, a cob structure and a traditional "Three Sisters" garden.

A Simple Plan

Developing a garden based on principles of ecological design can begin with effective, ecoliterate teaching activities. In our garden one upcoming project where we have consciously considered ecological design principles involves planting a simple grassland, or meadow, where a portion of the garden is planted with native grasses. Planting a meadow is a simple project that invites large-scale student experiential learning (many seeds to plant, many weeds to pull) while easily linking to interdisciplinary learning outcomes.

If a garden goal is to raise awareness of human EF, and to develop in students a love of nature and sense of eco-justice, while developing an appreciation for nature, the 'empty meadow' provides a powerful experience that also requires very little water. Honouring grassland also raises awareness of the ongoing destruction of that habitat for land development. Students in our garden learn about the symbolic importance of grassland in indigenous teaching and learning, how the grassland represents the earth and the honouring of 'space' that is not empty.

An Ecologically Just Cob Structure

Another simple, effective project that involved the class as a whole was the design and creation of a 'cob' structure for the garden. The assignment was simple: "Design a cob structure that is researched, planned and built in/as a biotic community."

Some ecoliterate design principles we considered before building the structure were as follows:

1. Preplanning considerations:

- What is there? What are we going to put there? How will nature help us in our purpose? (From Wendell Berry)

- What is our purpose in putting what we're going to put there?

- What elements in design/technique must be included in order to achieve that purpose?

- How can we work as a community to build the cob?

- What is the appropriate pace for this project so that we can

- a) leave a legacy for future learners and

- b) develop a mutually respectful community?

2. Design:
As you draw up your plans, reconsider responses to the questions above. Does your plan make ecological sense? Deconstruct the assignment as you go.

3. Process/Presentation:
Present your plan to the group.

- Be prepared to think of ways of incorporating elements from different designs.

- During this process did we achieve our 'purpose'?

- Did we remain true to ecological design principles?

The Three Sisters

A plant theme we recently worked on in the garden was "The Three Sisters", an ancient indigenous growing technique that includes corn, beans and squash. The three learning metaphors of our garden (environment, community and transformation) are reflected in this growing technique, as beans, corn and squash are planted together in mutual support.

According to Chief Red Arrow (2008),

> They are seen as the three beautiful sisters because they grow in the same mound in the garden. The Corn provides a ladder for the Bean Vine. They together give shade to the Squash. The Cherokee till the mound three times. The corn stalk provides a natural pole for the beans to climb. Their vines help to stabilize the corn stalk. The mature squash vines and leaves act as natural mulch, shading out weeds and holding in moisture. The beans fix nitrogen in their roots. If the plant is recycled into the soil, the nitrogen will feed the corn stalk for the following year. The spines of the squash vines are a deterrent to animals. The Three Sisters also complement each other nutritionally; corn provides carbohydrates, dried beans are rich in protein and the squash are rich in vitamins and minerals. The Three Sisters as a growing practice is based on traditional oral legends that vary among First Peoples. Before working on some lesson plans around the Three Sisters (see Activities, below) the students read the following legends.

Sky Woman: The Creation Story (Iroquois Legend)[1]

The term "Three Sisters" emerged from the Iroquois Creation myth. It was said that the earth began when "Sky Woman" who lived in the upper world peered through a hole in the sky and fell through the darkness towards an endless sea. The animals in the sea saw her coming, so they worked together and took the soil from the bottom of the sea and spread it onto the back of a giant sea turtle to provide a safe place for her to land. This "Turtle Island" is now what we call North America. Sky Woman had become pregnant before she fell. When she landed, she gave birth to a daughter. Sky Woman and her daughter lived on Turtle Island and when the daughter grew into a young woman, she also became pregnant. Legend has it that the West Wind impregnated her with twin boys but sadly she died while giving birth. Sky Woman buried her daughter in the "new earth" that was created for her by the animals of the sea. From her daughter's grave grew three plants that became sacred to the land—corn, beans and squash. These plants provided food for Sky Woman's grandsons, and later, for all of humanity. As time went on, these special gifts ensured the survival of the Iroquois people. The three sacred plants came to be referred to as the "Three Sisters" and according to the Iroquois people, if you grow the Three Sisters you will never go hungry. The legend of the Three Sisters varies across aboriginal groups, yet the sacred value of corns, beans and squash remains consistent.

The Three Sisters (Mohawk Legend)[2]

A long time ago there were three sisters who lived together in a field. These sisters were quite different from one another in their size and way of dressing. The little sister was so young that she could only crawl at first, and she was dressed in green. The second sister wore a bright yellow dress, and she had a way of running off by herself when the sun shone and the soft wind blew in her face. The third was the eldest sister, standing always very straight and tall above the other sisters and trying to protect them. She wore a pale green shawl, and she had long, yellow hair that tossed about her head in the breeze. In one way the sisters were all alike, though. They loved each other dearly, and they always stayed together. This made them very strong. One day a stranger came to the field of the three sisters—a Mohawk boy. He talked to the birds and other animals—this caught the attention of the three sisters. Late that summer, the youngest and smallest sister disappeared. Her sisters were sad. Again the Mohawk boy came to the field to gather reeds at the water's edge. The two sisters who were left watched his moccasin trail, and that night the second sister—the one in the yellow dress—disappeared as well. Now the elder sister was the only one left. She continued to stand tall in her field.

When the Mohawk boy saw that she missed her sisters, he brought them all back together and they became stronger together, again.

The Teacher's Role in Developing Eco-Centred Curriculum

The above sample activities (the meadow, the cob structure, the Three Sisters legends) are meant to be taught through mindful hands-on/minds-on involvement. Truly eco-centred activities require ways of teaching and learning that seem slower, community based, and process oriented. These ideas coincide with ideals around 'the slow school' movement, described by Maurice Holt (2005) as

> a metaphor for a particular approach to practical problems. Hence the idea of a 'slow school', which doesn't mean reading in slow motion for slow learners. The slow school attends to philosophy, to tradition, to community, to moral choices. Students have time not just to memorize, but also to understand. (p.60)

In a garden setting, what is the teacher's role? Even as a teacher facilitates the flow of information, an outdoor classroom dramatically alters the role of the instructor, as learning begins as a collective community. Developing deeper lessons in ecology in a school garden can be a slow pedagogical process. How can a teacher facilitate critical learning that allows students to openly and spontaneously honour the natural world?

One teacher educator (Krapfel 1999) developed a project that included 10 curricular units that helped teachers investigate natural elements (i.e., birds, ants, trees) specific to their regions, and to their school grounds. In visiting schools that adopted this 'school ground science' curricula, he noted how, with universal, hands-on access to the projects on the school grounds, students from across grades were teaching each other. In one technique, he had students measure the growth of a flower. Students marked their flowers with lightly tied string, then noted changes. The students went outside and graphed their findings: *Has the flower lost its petals? Has the rate of sunlight and shadow affected the growth of the flower?* The instructor noticed how the activity helped students slow down and notice specific elements, generating a slower, more careful awareness of the rate of change in nature.

Another school instructor describes how he gets students to really look at their flowers:

When we did our sketching, I told the children to pretend they were magnifying glasses: 'Actually look, *look* at all the little things. When you draw a flower, it isn't a kindergarten flower any more. Take a look at it. It's got veins, and hair, and little things all over it.' They really were silent, they looked, they watched and they came back with some beautiful detailed drawings. But if they hadn't thought about it...they would have had a different way of seeing it. (Houghton 2005, p.77)

The teacher educator (Krapfel 1999) goes on to describe how, when he first started the programme, he required that students achieve uniform results, and note only certain aspects of growth, in order to achieve standardized, measurable outcomes. He states how this method led him into 'bad teaching' (p.51), as the students were bored with homogenized results that left out variables, or even surprise. He realized how his previous training in science had taught him to seek certain outcomes, while ignoring the larger context of both the natural world and authentic, imperfect, student participation.

The instructor describes his own transformation:

In wanting to focus on one variable that would teach a chosen concept, I was discounting the effects of all the other variables. I was discounting the opportunity for students to directly experience how complexly interconnected the world really is. This happened enough times for me to begin loosening up. Rather than jumping in to discount 'extraneous' variation, I found myself holding back and nourishing group discussions about what the results meant. This change began leading to class discussions concerning all the variables that might be shaping the results and the different possible interpretations that could explain them. (p.51)

As we similarly discovered in our campus garden, the instructor noted how simply being in an outdoor space challenged the hierarchies of teaching and learning:

The classroom is a teacher-controlled environment. Which subjects are taught when, which activities are done, and what is marked correct or incorrect are controlled by the teacher. But kids realize that they cannot control what is happening in the field. A teacher is expected to know the answer to every question asked in a textbook, but students do not expect a teacher to know everything in the field. The entire focus shifts from question-answer dialogue to sharing and wondering...Rather than the teacher examining the student, they both stand together examining the world in wonder...The important point is that kids respond better to the real thing with all its complexity than to simplification. (pp.52–53)

There are a variety of ways researchers suggest will bring students closer to 'the real thing' through curriculum that invites complexity, creativity and critical knowledge. Rachel Carson describes the important role of 'adult interpreter' of nature (cited in Corcoran 1999, p.181); meanwhile, Edmundson advocates designing curriculum that leads students to understand urgent, complex issues now involving the planet. In his paper *Rethinking Food: Towards a Curriculum for Sustainable Food* (2006) Edmundson suggests designing curriculum around the following principles:

• Direct food (growing food/establishing school gardens);

• What kids eat (changing students' food habits i.e. organic produce from gardens vs. vending machines);

• What kids learn (developing challenging curriculum around globalized food and its effects on consumers, workers, and farmers and on economies of third world; resisting corporate influence in schools, genetically modified food; effects of pesticides & fertilizers);

• Challenging how food is grown, processed & delivered;

• Challenging globalized, corporate food systems that dominate small, local farms. Challenging the culture of modernity (challenging the concepts of individualism and anthropocentrism). (pp.3-4)

Edmundson suggests that in ecology-based curriculum development, discussions of environment and food need to include both critique and action around a wide range of disciplines. Meanwhile, the Toronto District School Board uses three broad concepts in integrating environment to established curriculum:

> Sense of Place: to explore and see our immediate surroundings, both natural and built, and to look beyond the immediate to the larger landscape. Ecosystems Thinking: to help us examine how nature works and to understand how natural and human systems are interconnected and interdependent. Human Impact: to weigh the consequences, both helpful and harmful, of human interventions into natural systems. (Cited in Houghton 2005, pp.20-21)

The BC Ministry of Education Curriculum Supplement *Environmental Learning and Experience, an Interdisciplinary Guide for Teachers* (2007) suggests designing curriculum around the following principles:

- Encourage the integration of subjects/multidisciplinary approaches;

- Encourage critical reflection on a range of perspectives;

- Examine issues for their currency and authenticity;

- Acknowledge aboriginal perspectives;

- Acknowledge other perspectives;

- Consider the place of action;

- Consider issues from both local and global perspectives;

- Occur within a context of hope;

- Encourage humility.

Both of these new curriculum design guides do not so much focus on 'how to' models of 'inclusion' around environment, but instead emphasize an approach where learning literally 'takes place' within local communities. These new curriculum documents, in their use of words like 'hope', 'humility', 'interconnected and interdependent', are strongly influenced by global environmental research from the UN that emphasizes how the effective integration of ecology in schooling must begin with, and retain, high ideals, broad visions and hopeful perspectives.

As for evaluation, in ecological and community models of learning, learners largely evaluate and peer review their own learning pathways, based on guiding principles that ultimately support and sustain life and natural processes. As described extensively in this text, eco-centred learning supports wholeness; 'success' in such a setting thus depends on long-term survival and inclusivity in a community that is supportive of alternatives and variations. Students are very much a part of the whole evaluation of the project and are encouraged to set up evaluation criteria for projects that bring about sustainability and renewal of the natural world.

In our garden, stewards were wary of self-glorification in developing 'knowledge' in the garden and developing tightly defined learning outcomes that would restrict student learning. Holistic assessment for projects in the garden involved questions like How does this project help sustain the garden? Does the project preclude or deny others' involvement or ideas? Learning in a garden involves give and take, and thus evaluation involves the development of holistic attributes in teaching and learning. For example, students develop flexibility and non-competitive modes of learning (and they learn sometimes that even their 'good' ideas might not work in the context of the garden/the community). Students develop listening skills and interpersonal skills based around values of respect, equality and peace. The ultimate 'evaluation' of any eco-centred learning project always returns to the land: how does this project help sustain natural life?

As we also learned, the teacher's role in learning in a 'school garden' involves both guiding and facilitating knowledge and allowing progressive, interdisciplinary practice to take place. In considering how the garden could promote interdisciplinary learning, social and team skills, PBL, community access and preventative classroom 'management', we realized that the garden truly supports so many progressive ideals in practice. Similarly, the idea of teaching and learning in abstraction, using single subject models, is virtually non-existent in a garden. If you are, say, examining, discussing, writing about or plotting out a bed of strawberry plants, it is impossible to classify the work as science, art, math, home economics, social studies, English. A garden promotes truly eco-centred learning where the land, sustainable practice and community mindedness are seamlessly integrated. In a garden, the teacher is not an actor pulling the strings of the students in attempting to generate interest or even enthusiasm. Instead, eco-centred learning in a garden involves developing a plan that will shed light on student understanding of the garden, the larger community and ecosystem surrounding the garden and, ultimately, to the learning process itself.

As we learned in our campus garden, a garden is not just 'a garden', but a place that invites connections for learning. While learning begins in the garden 'as environment' the effects and the impact on teaching and learning can be far reaching. Grant and Littlejohn (2004), in *Teaching Green, the Middle Years*, discuss how the practicalities of 'teaching green' actually opens a teaching and learning environment to even more profound changes in conceptual thought, practical action and learning objectives. They describe learning goals around environment through a diverse range of characteristics where, for example,

• Students should have opportunities to develop a personal connection with nature;

- Education should emphasize our connections with other people and other species, and between human activities and planetary systems;

- Education should help students move from awareness to knowledge to action;

- Learning should extend into the community;

- Learning should be 'hands-on';

- Education should integrate subject disciplines;

- Education should be future-oriented;

- Education should include media literacy;

- Education should include traditional knowledge. (pp.xv–xvi)

In describing the role of teachers as "facilitators and co-learners" (pp.xv–xvi) these researchers ultimately make a simple, effective point:

> Teachers do not need to be experts to teach about the environment. The natural world is an open book for endless discovery by all. As co-learners alongside their students, teachers both model and share in the joy of learning. (p.xvi)

Other researchers note how, in a school garden, the 'hidden curriculum' becomes a central part of the learning process:

> Sometimes hidden curriculum is intentional: in CEL terms, changing the context from the four walls of the classroom to include the culture and the community of the school, experiences in the garden, faculty talking to one another respectfully, or the use of a tablecloth in the dining room at the Edible Schoolyard are all hidden curriculum. (Callenbach 2005, p.45)

Too often, teaching *about* the land does not involve learning *within* the land. The teacher's role, in addition to facilitating ecoliterate starting points for curriculum, is thus also about modeling the ideals of nurturing community through the land, of working alongside students and teachers in community.

Before the establishment of models for eco-centred and sustainable learning, students and teachers must first locate nature. In fact, the first step in connecting students and teachers to nature is learning 'what' and 'where' nature is. The next step is to go outside. Such a move, like the building of a school garden, is a small but important symbol in helping students learn about nature from within nature, in helping students learn about ecological systems within the actual experience of those systems.

School Gardens: Practical Matters

While there is no perfect recipe for the practical development of a school garden, experienced educators describe how certain practices can help nurture meaningful teaching and learning in outdoor settings.

School Support and Small Steps

A garden begins in community, and in nurturing support for an idea that may, at first, seem a bit unrealistic. Hope and enthusiasm are important starting points in developing support. Neil Smith, former principal of Martin Luther King Middle School in Berkeley, California, explains:

> First of all, you need to understand that it's not going to happen in a year, that it's a long-term project. Second, you don't have to have buy-in from the whole staff initially to get started, but you do have to have a few committed people. Third, it's good to look for support both inside and outside the school, at least initially, to make it work. You need district support. Our superintendent, Jack McLaughlin, was very supportive, and this is especially important if you are going to be using part of the land at the school. Finally, there has to be support for the teachers who participate, support in various ways like time to meet or a person who is out there guiding them. (Cited in Comnes 2005, pp.146–147)

Renay Weissman adds to Smith's advice, noting

> I would not advise anyone to initiate a project without support from the principal. You must have support from the administration. (Cited in Houghton 2005, p.66)

UN research also indicates that school-wide support is vital in sustaining school gardens:

The garden must be viewed as an integral part of the educational plan for the school (e.g., as a classroom) and financed accordingly as a part of the overhead of operations. If this is not the case, then long-term sustainability is in jeopardy and the garden becomes a burden to the creative energies of staff, parents and community volunteers. (Desmond, Grieshop & Subramaniam 2004, p.71)

Miriam Mutton and Debbie Smith (2001) recommend starting a school garden with a variety of school and community support groups:

- Form a steering committee consisting of a few people with a common goal and lots of energy to get the project wheels turning. Include parents, teachers, student representatives, the school grounds caretaker, and a school administrator.

- Establish a regular meeting schedule to keep the lines of communication open and clear. Let the frequency of meetings be determined by the stage of project development.

- Set up a system to maintain communication within the group, school, and community. Newsletters sent home with students and a letter sent to neighbors will keep parents and local residents informed. Other avenues of communications include the public relations person at the board office and the local media. (p.25)

Community Involvement

Starting a garden directly involves community:

Reaching out to discover community resources is a valuable first step in getting started. Some would-be gardeners begin by visiting other school gardeners and their gardens. Others gather ideas by visiting parks, nurseries, conservation authorities, outdoor education centers, arboretums or local natural spots such as ravines. Such preliminary visits often blossom into ongoing cordial relationships and exchanges between schools and local plant experts. Once the idea of a garden is set into motion, schools might discover that there are experienced gardeners or even landscape architects right in their midst, among their own staff and parents. (Houghton 2005, p.31)

The idea of direct involvement and need for community participation in school garden initiatives is echoed by Stephanie Stowell (2001):

The initial spark that sets an outdoor classroom project in motion may come from one inspired teacher, parent, or community member. However, for the spark to be sustained and the project to succeed, there must be leadership from a wide variety of stakeholders. The project team is the working committee that oversees the development and implementation of an outdoor classroom. Their tasks are to coordinate site planning, fundraising, publicity, building, planting, and maintenance of the diverse skills and support brought by each member of the project team are invaluable in the overall conception, construction and maintenance of a project. To ensure a solid foundation, a project team should include representatives from several key groups. (p.16)

Developing a vision, designing a plan, deciding involvement are important parts of the process that should unfold organically. From the beginning, 'the goal' should involve the process. In our garden, for example, we discovered how our initial weeding of the area turned into an opportunity for community building.

Neil Smith describes how revitalizing the soil became a community event:

even though we got rid of the asphalt, the soil was not good. So, in December of 1995, we had a planting. We invited people connected with the project to come to throw the first seeds. We made it ceremonial, with an Aztec dance group. They were drumming while we got lines of people to walk along and toss seeds as they went. The seeds were primarily fava beans, to put nitrogen back into the soil and to cleanse the soil. (Cited in Comnes 2005, p.142)

Enthusiasm for Learning: The Middle School Connection

Middle school is in many ways an ideal age group for teaching and learning in school gardens (Grant & Littlejohn 2004). *This We Believe: Successful Schools for Young Adolescents* (National Middle School Association 1995) defines middle school philosophy and practice through certain 'characteristics' of middle school that include

- an inviting, supportive, and safe environment;
- students and teachers engaged in active learning;
- multiple learning and teaching approaches that respond to their diversity.

Methods for achieving such characteristics include collaborative planning, team teaching, cooperative learning, experiential learning and community involvement. Characteristics and methods for an 'ideal' middle school thus coincide well with the principles of teaching in school gardens.

As principal Neil Smith describes,

> One of the tasks of middle school is to excite kids about learning, and the Edible Schoolyard is successful there. If at the end of a child's eighth-grade experience, the child has certain facts down, how long will those facts be remembered? But if the child is excited about a particular subject, he or she is going to become a real student and go on learning about it. That's really what you want to do—in middle school in particular—isn't it? (Cited in Comnes 2005, p.148)

Teacher Michelle Barraclough gives a personal anecdote about the hands-on nature of her school's garden, which illustrates to the students a focus on learning engagement:

> I know with my grade ten class, when we started off, it was 'Do we have to do this, Miss?' noted Barraclough. 'What I do with my grade ten is I give them a plant', and I say, 'This is your plant.' They start to look after it, and then it's 'Is my plant okay, Miss?' (Cited in Houghton 2005, p.69)

Finding Themes: Cycles across the Curriculum

Fritjof Capra, founding director of the CEL in Berkeley, California, describes how an entry point for teaching in school gardens involves teaching *in* and *as* an ecological system:

> When school gardens are made part of the curriculum, for instance, we learn about food cycles, and we integrate the natural food cycles into our cycles of planting, growing, harvesting, composting, and recycling. Through this practice, we also learn that the garden as a whole is embedded in larger systems that are again living networks with their own cycles. The food cycles intersect with these larger cycles—the water cycle, the cycle of the seasons, and so on— all of which are links in the planetary web of life. (Capra 2005a, pp.xiv-xv)

In practice, at Silveira school in San Raphael, California, students become an actual part of a learning cycle, through which they learn about natural cycles. Principal Jeanne Casella explains how she incorporated the idea of 'cycles across the curriculum':

> We chose 'cycles' as an ecological concept around which to integrate curriculum for the year. I used several resources to introduce the concept. Then

teachers examined the curriculum and decided how to weave the study of cycles into it. For instance, while working in the garden, first-graders studied the life cycles of beneficial garden insects such as the ladybug and the praying mantis. Students working on compost studied nature's recycling processes. (Cited in Barlow, Marcellino & Stone 2005, pp.155–156)

In this setting, students work at 'jobs' on 1 of 22 teams that look after the school grounds and community:

A pond team watches over the pond. Junior master gardeners assist in the garden. The energy team conducts energy audits and writes notes to teachers, providing gentle reminders to turn off the lights when a class goes to the cafeteria. (p.149)

Start Small and Build

A school garden can start small, with a meadow or a single box. If summer watering is problematic, a garden can begin with simple observation of a site.

Planning, of course, is a key component of all successful gardening. For those who wish to plant discrete gardens, an essential up-front planning consideration is maintenance during the summer, the heart of the gardening season. Provision must be made for weeding and watering through the months when schools are not in session. For beginners a discrete garden is a practical place to start, if we bear in mind the advice to "start small" offered by many successful school gardeners. A discrete garden is a manageable space to measure, plan, plant, nurture and maintain. Yet gardening on this small scale in no way limits the degree of creativity that can go into the garden's design. (Houghton 2005, p.46)

Another way of 'starting small' involves creating a three-dimensional model of your proposed gardening site prior to its physical development. There are several benefits to this:

Most people are visual learners and have · trouble translating a two-dimensional site plan into a three-dimensional space and vice versa. Making a large, three-dimensional model of your school grounds will enable you to visualize the project as you develop a plan for the site. A model will also make it easy to explore and evaluate the various options you are considering. By making movable to-scale replicas of components such as fences, pathways, and plantings, you can determine the best placement of the various elements you have in mind, avoiding pitfalls such as placing conflicting activities side by side, obstructing sight lines, and locating projects across 'desire lines' (non-designated pathways that people choose to use to get from one place to another). A model can be invaluable in generating excitement for the project among the school community, in gaining the support of school board planners, and in raising funds. Integrating model-making into the curriculum immediately connects children with every part of their school grounds and helps to generate a sense of ownership in the project. (Coffey 2001, p.29)

Students could also start small by 'fallowing' the site, or stripping the site of green lawn/asphalt and observe the area for one season. As Arthur Beauregard, manager of Natural Environments and Horticulture for the Parks and Recreation Department of the City of Toronto, notes,

> There can be a whole education in just seeing what's growing, noticing for example if it's an annual weed or perennial weed growing from seed. Then you could have the opportunity to find out how plants grow. The first season should be taken up with actually cutting the annual weeds down, or pulling them up before they go to seed, and in digging out the complete root system of the perennial weeds. (Coffey 2001)

After the first year of observation, Beauregard suggests planting in shallow areas of sand:

> [sand] suppresses the germination of any weed seeds that still happen to be on the surface...the other advantage to sand is, if you're growing native plants, a two-centimetre layer of sand will markedly cut down—by a factor of about 95 percent—the number of weed seeds. (Cited in Houghton 2005, p.52)

Organization/Roles

Many experienced school gardeners advocate a community approach that involves students, educators, administration, maintenance staff and parents.

One teacher (Stowell 2001) recommends involving student creativity and leadership in developing the vision for the garden by asking them questions: "How do they currently use the schoolyard? How would they like to use it in the future? What features would they like to have in an indoor classroom?" She describes the teacher's role:

> As the individuals who are most directly involved with students, teachers are an integral part of any successful project team, serving as mentors, resource aids, facilitators, co-workers, and cheerleaders. They must strike the balance

between taking the lead to push the project forward and stepping back to allow for student initiative. (p.16)

Other researchers discuss the importance of including community directly in the organization of the garden:

Many school garden programmes have had success with students working in small groups on different tasks. Inviting knowledgeable parents or volunteers from the community to supervise small groups of students, or even using older students as mentors will help make this possible. (Welby 2003, p.1)

Stowell (2001) also notes the importance of community involvement in school garden projects:

It is essential to include members of the wider community as key players on the project team. Natural resource professionals can offer expertise that may be difficult to find elsewhere, and local businesses and civic organizations can offer technical support, in-kind contributions of materials, volunteers, and, perhaps, grants. You may find assistance from a wide variety of sources such as landscape architects ready for a new challenge, local businesses willing to donate plants and other materials, or garden and civic clubs excited to offer their knowledge and hands-on assistance. (p.18)

More Tips

- Make expectations about outdoor behaviour clear. Establish guide-lines about proper use and care of tools, organization of materials, where to walk in the garden, and what to do if finished early.

- If your group will often be working outdoors in the garden, create a meeting space with places to sit and some shelter from wind and rain to help students stay focused during whole class discussions. (Welby 2003, p.1)

- Use spontaneous moments outside to your advantage. If a rabbit hops along the trail while you are teaching and students turn their atten-tion to it, take the broader view of learning and turn these opportuni-ties to your advantage.... (McCutcheon & Swanson 2001, p.125)

- Carry a prop bag with different nature items in it to refocus students when you want their attention again. The suspense of what your next item will be will help get their attention back. Items such as nature bingo, scavenger hunts, seeds, leaves, feathers, etc. can be things you have in your bag. (p.126)

Activities

There are a steadily increasing number of activities and learning resources available around school gardens and eco-centred learning. The following list of activities, some developed in the garden working with student teachers and some sourced elsewhere, is meant as a starting point for teachers to integrate further lessons in ecology and school gardens. (Also see 'Teacher Resources,' below).

ACTIVITY: Three Sisters

The student teachers in the garden explored the history of the growing practice and created the following garden-based activities for middle school students.

1. Collaborative Poem

As a three-person team, have students create an oral poem around the Three Sisters. Students will write the poem as if they were alive during the time described in the Iroquois or Mohawk Three Sister legends. [3]

(Suggestions) Have students brainstorm the following ideas:
• The importance of the Three Sisters to yourself and/or your family
• The meals that you prepared using the Three Sisters
• Your journey in growing/gathering the Three Sisters
• The plants that grow around the Three Sisters.

2. Poetry in the Garden: Composition Inspired by the Three Sisters Agricultural Method

Each student will write one stanza of the poem inspired by one crop from their Three Sisters garden and then combine their poem with two other students to form a three-stanza poem. Modeled on the Three Sisters agricultural method, the lesson highlights the value of community and diversity within the creation of poetry. [4]

3. Three Sisters Wall Poem

Divide students into small groups. Each group will choose one of three themes/topics: seasons, people or plants/food. Students will write short lines of poetry (4–5 words per line) which relate to their theme/topic (i.e., season, people or plants/food) and to the idea of the Three Sisters. Each group will be given paper cut-outs of corn husks, bean or squash leaves on which to write their poetry lines. The 'season themed' groups will be given the squash leaf shaped paper cut-outs; the 'people themed' groups will be given the bean leaf shaped paper cut-outs and the plants/food themed groups will be given the corn husk shaped paper cut-outs. These cut-outs will be placed on the poster in the appropriate areas to create a giant visual class poem about the Three Sisters.[5]

4. Just Plant

This activity takes place in a school garden. In groups of three, students plant the "Three Sisters". Similar to the Three Sisters, the students in their groups of three will work together to maximize the growing potential of their Three Sister plants. This activity is hands-on. Students will be given corn, bean and squash seeds, gardening spades and watering cans in order to perform this activity. Working in the garden with their three plants (in groups of three) will provide students with firsthand understanding of the relationship between the sisters.[6]

5. Three Sisters Support System

Read the stories of the Three Sisters to the students in the garden. Invite the students to bring their attention to the sounds around them. Encourage them to think of two people in their lives that they have supported, and who have also supported them. Suggest that they seek inspiration in friends, family or other people in their community. Shop owners, bus drivers, mail carriers and others who provide businesses or services could all serve as inspiration.[7]

- Give the students a few minutes to visualize the ways in which their trio interacts and creates positive experiences for all.

- Students will create their own legend about the ways people from their own lives inspire and demonstrate interdependence of all life, and the way the actions of one can impact others.

- Students develop a 2–3 minute oral story to recite, act out, or sing for their classmates around the Fire Pit at the garden. Include drums and rattles, as well as an assortment of costumes, in their presentations.

ACTIVITY

Walk around the school grounds/neighbourhoods and have students write poetry, draw maps or research their local habitat.

ACTIVITY

Develop a 'garden blog' where daily activities, student work and changes in the garden are documented through writing, photos and film. Incorporate garden blog participation 'across the curriculum' into course evaluation. Share the blog with other 'green campus' networks or other national/international schools. Share the blog, writing, garden ideas and photos on social networking sites.

ACTIVITY

Develop a local school garden network—set up field trips to other school gardens. Set up an assembly where students share their work/photos in the garden with the entire school.

ACTIVITY

Hold a 'harvest sale' or 'fund raising fair' for your garden in the fall. Invite local musicians from the school, high school or local college.

ACTIVITY

Have students develop an ecological design plan for an irrigation system. Students will research the plan and develop a report for a school garden committee. Students could research standard and alternative forms of irrigation (i.e., gravitational means, a rain barrel or traditional irrigation methods learned from the local aboriginal community). As a group, have students assess the worthiness of the plan based on a set of criteria that will ensure sustainability and responsible water use. Have students research worst case and best case scenarios.

ACTIVITY

Reach out to the community. For example, initiate an intergenerational learning project, where students learn in the garden alongside elder 'buddies' from the community.

ACTIVITY

Donate produce from the garden to the local food bank. Encourage student involvement in delivering the produce; have students report back to class on their awareness of their local community.

ACTIVITY

Partner with the school cafeteria to create an organic composter and/or donate or sell garden produce. (Encourage the cafeteria to sell organic food.)

ACTIVITY

If you are at a school, try partnering with the local college and invite students and instructors to visit your garden; if you are at a post-secondary college, invite local students/seniors/community groups to visit the garden.

ACTIVITY

Hold community volunteer days at your garden and invite locals through an advertisement in the local newspaper or through local community garden Web sites. Invite a local newspaper or radio station to volunteer day.

ACTIVITY

Invite students to write about the 'history of their family's interaction with the land'. Where did their family originate? Did any of the students' ancestors farm? Are there any descriptions or photos of those traditional techniques? Have students display visual and written narratives around the classroom.

ACTIVITY

Ask students to write about their 'very first memories of nature'. Was it a farm? An apartment patio window box? A summer adventure?

ACTIVITY

Invite students to make a list of places in nature that are important to them. Have them share the list with a partner; then select one place to write about. Ask students to finish the sentence: "Whenever I think of (the place) I think of. . . " Develop poems around the writing. Copy the poems on construction paper. On the other side of the paper, have students create 'earth logos' that visually symbolize their values and hopes around the land and sustainability. Using sustainable materials, make flags out of the paper and have the students walk around the campus grounds (or to the garden) to read their poems aloud.

ACTIVITY

Using donated or reclaimed cloth, have the class stitch a shade canopy for the garden supported by four poles made of organic materials.

ACTIVITY

Designate a tree on campus as a 'Peace Tree'.
Hold a class discussion around the principles of peace, ecology and global justice. On a small piece of recycled paper, write the words 'Peace Tree', the year and the grade level of the class. Have students sign their names on the paper, in support of peaceful principles. Walk to the Peace Tree and, using earth friendly string, ask the tallest student in the class to tie the string around the highest branch they can reach. Ask every student to say aloud their 'wish' for the earth. Conclude with a song or a poem. Repeat annually. (Invite students to return to visit the tree in future.)

ACTIVITY

Invite students to cook a meal slowly (for younger students, have someone help them cook a meal) where every ingredient is considered in terms of its original EF. Have students report back on their findings. On a wall map, attach pins in the regions where their dinner originated.

ACTIVITY

Invite inclusiveness in your garden; invite a member of the local indigenous community to consult on the development of the garden. What are the best plants for this region? What are the plants that the students can really learn from? What learning practices can be incorporated through discussion with local bands and elders? Does the indigenous community have a symbol (i.e., a carving) that will help the students focus their learning on a specific place or certain symbols?

ACTIVITY

Support oral language and local craft in the classroom. Invite a storyteller to tell stories in the garden while students make pine baskets or some other craft based on local, reusable materials.

ACTIVITY

While telling and listening to oral stories, either live or recorded, have students make lavender 'wands', small sachets or other crafts for a fundraising sale.

ACTIVITY

Create a rotating model of garden stewardship, involving different students each year. Allow fresh ideas to cultivate.

ACTIVITY

Encourage students to find recycled and/or donated parts to build planters, benches, etc. The search may lead to the teachers discovering some interesting grass-roots local networks, such as the world of recycled bottle scavenging, the metal scrap collection community, etc.

ACTIVITY

Create social inclusivity around the garden. Encourage students to speak to plant facilitators and maintenance workers on campus for social discussion, to enlist their help or to find out about upcoming projects.

ACTIVITY

Take time to develop a plan for a garden (and a classroom) that reflects the needs of local place and community. The first question for GBL is always How does this project positively add to the natural environment? To the larger ecological community? Does this project include the involvement of everyone in the garden community? This process should take time and discussion.

ACTIVITY

Take a trip to the local farmer's market. If possible, give groups of students a small amount of cash and have them purchase an item from a farmer. Ask students to discuss with the farmer one practical farming technique or concept that could be applied to their garden practice. Over lunch, discuss the ecosystem of land, farmers, local food and how the class can actively support that system.

ACTIVITY

Brainstorm a list of 'out of the box' ideas for the garden. (In the Learning Garden, this is how we came up with several innovative projects, including a proposed 'cob' structure.) How about a star gazing area? An empty meadow? A place for the school/campus staff to relax? A 'sensory' area? Garden beds planted in the shape of the alphabet? Planting a collection of plants that appear in children's stories? Shakespeare plays?

How about inviting teachers, administrators, even local officials to recite poems or stories about nature? Does the mayor or other city councillors have a special poem to share with the community?

ACTIVITY

Build a 'cob' tool shed, bench, chair or wall made of sand, clay and straw. As the experts at 'Cobworks' explain, the materials must be

> wet enough to mould yet dry enough to build up without forms. Walls are built up to be monolithic, which gives them greater strength in earthquakes. Cob invites your creativity to be expressed in the process of building your space: living space, meditation space, work space, sleeping space, play-space. Build curved walls, arches, and niches. And who says that walls are flat? All it takes to make the step towards your dream space is a fairly short learning process, to gain a sense for the material. A fun way of learning is to participate in a workshop, where you work on a structure and learn hands-on, with opportunity to discuss your ideas with others, pick up knowledge of the whole process and meet new friends. [8]

ACTIVITY

Listen to the trees.

Describe the sounds, one by one. Later, include one remembered sound for each line in a poem.

ACTIVITY

Take a field trip to the nearest body of water. Have students make lists about the body of water's ecological/biotic characteristics. Ask students to write about the sensory characteristics about the water. Have students develop poems around the body of water.

Try developing metaphors around the water and finish the following sentence: "This body of water sounds like...smells like...looks like..."

Invite students to enter the "River of Words" contest founded by poet Robert Hass. (www.riverofwords.org)

Hold a school-wide poetry reading on 'bodies of water' and screen a film around the current state of the world's waters.

ACTIVITY

Write poetic 'odes' to the Three Sisters. Divide students into three groups: one group will write about the corn, one group will write about the beans and one group will write about the squash. Gather around the Three Sisters plants and recite the poems, in threes.

ACTIVITY

Help students connect with land as a 'bioregion', or as a geographical region that extends beyond human organizational or political borders. Invites students to map out their 'bioregion' by: listing the native plants in the region; identifying bodies of water and other geographical features (i.e., mountains, valleys); building pathways in and around the school garden that link geographies and connect the garden to the surrounding area.

ACTIVITY

Have students keep a garden journal. This is an important component of learning in the garden that allows students to reflect on their experiences. You may want to give your students specific writing or drawing tasks for their journal, or simply set aside time for students to write unstructured responses. Journals can also be used as an assessment tool.[9]

ACTIVITY

Divide a garden study area into 1 square meter plots using string and stakes. Divide the site into smaller grid units by marking the sides at regular intervals. In the field, have students record details of date, time, compass or GPS (Gallons Per Second) location, weather conditions and temperature. Have students sketch a scale map of the site on graph paper, adding letters and numbers along the two axes to provide coordinates. Have them record the locations of plants, trees, rocks, shrubs, water, bare ground, etc. Let students count and identify all the plants in the plot, take measurements and make drawings. Once back in the classroom, make a master list of plant species between all student sites. [10]

ACTIVITY

• Make garden journals to keep throughout the year. Provide time to write, draw, record and paste up after each garden time.

• Make a scarecrow. Hammer two narrow boards in a cross. Ask students to bring child-size shirt, pants, shoes, mittens and accessories. Head can be an old t-shirt stuffed and rubber banded.

• Build a school community compost area and start composting the vegetable and fruit scraps from lunches.

• Make recycled paper and add flower petals.[11]

ACTIVITY

• Plant a pizza garden.
• Design an ABC garden, which includes an example of a native plant beginning with each letter of the alphabet. [12]

ACTIVITY

- Take students outside and have them collect information about the flora and fauna they find, then use computer, art and science skills to work to make a field guide. You can use a computer drawing program, pencil crayons, fine-tipped markers, pens, clipboards, journals, paper, a jar for collecting insects, magnifiers, a camera and film to record and journal the students' findings of the surrounding schoolyard's ecosystem.

- Show students how to set up a database. Make a standard form for recording information. Data fields include the common name for each flora/fauna entry, scientific name, family, name of discoverer (student's name), date, location, habitat and unusual features. Separate records for the plant, animal and insect kingdoms and for geology. [13]

ACTIVITY

- Use the natural area for reading, either silently or aloud.

- Have students write stories about how to reduce our EF, using the natural area as inspiration.

- Use the outdoor classroom as inspiration for poetry that is regularly added to a "poet tree" in a central location in the school. [14]

ACTIVITY

- Chart sun and shade patterns over time to determine where to put various plants in order to meet their specific needs.

- Use the outdoor classroom as a tangible place to apply math concepts such as perimeter, circumference, diameter, volume, angle, ratio and scale. For example, students can measure diameter, circumference and ratios of plants and trees; chart and graph growth rates in plants; estimate changes in abundance and diversity of insect populations; calculate and compare the volumes of water required by various plants. [15]

ACTIVITY

Conduct environmental assessments that test soil, air, pollution levels, soil compacting and so on. These studies can be extended to different environments around the community. Students may investigate questions such as how soil conditions, the pH of the soil or the availability of water and sunlight affect plant growth and the presence of animals.[16]

ACTIVITY

• Write songs about the development of the nature area, or use the area and environmental themes as inspiration for song lyrics.

• Make "orchestra cards" that use pictures to describe natural sounds that students hear in the schoolyard.[17]

ACTIVITY (Save the Turtles)

Have students investigate a local environmental issue and address it through the school garden. If wildlife habitat is being destroyed, have the students raise awareness by creating art projects such as flags and stringing them up around the garden. Invite the local newspaper to discuss the issue.

ACTIVITIES (Calculating EF)

The following Web site can help students figure out their EFs. The site is user friendly, and not too technical. All of them could be used at the start of the year, throughout the year and at the end of the year to see the rate of improvement based on action taken by students to green their lifestyles.
• http://footprint.wwf.org.uk/

The WWF Footprint Calculator is a quick, simple Web site you can use with students to calculate their EFs. It takes about five minutes to complete the survey, and gives you your EF total in carbon footprint tones per year, as well as how many of our planets it takes to support your lifestyle, along with percentages of what various areas of your lifestyle are consuming. Suggestions are provided at the end of the survey for how to live a more eco-friendly lifestyle. It is UK based, so money is in Pounds and there are a few other specific British references.
• http://www.epa.vic.gov.au/ecologicalfootprint/calculators/personal/intro duction.asp

An Australian EF calculator that takes a bit more time than the one above. Includes more in-depth questions and responses as well, giving a more accurate representation of your EF. The site includes EF calculators for Personal, Home, School, Office, Retail and Event use, and also calculates EF by how many of our planets are needed to support your lifestyle. Once the quiz is finished, suggestions for energy reduction are presented, and when clicked, show how much your consumption can be reduced by. The quiz is accompanied by a personal avatar, engaging graphics and sound. The site has visual appeal.

• http://www.myfootprint.org/en/

The Ecological Footprint Quiz estimates your EF based on a 10–15 minute quiz. It is quite in-depth as well, providing a more accurate rating of your EF. Your country's average is compared with yours throughout the quiz, so you can see each response affecting your total based on your choices. Younger students might not know the answers to some questions, such as the square footage of the house they live in, or the number of kilometres they have travelled by plane in the last year, so younger students might need some help on some questions. EFs are presented by the number of earths needed to sustain your lifestyle if everyone consumed/emitted as much as you, and your EF is compared to the national average in charts, tables and graphs. Suggestions are offered at the end of the quiz to help you reduce your EF.

Teacher Resources

The following Internet sites are a small sample of the growing Web resources available on school gardens and ecology-based learning.

Internet Resources

BC Agriculture in the Classroom Foundation
http://www.aitc.ca/bc/
Provides resources and lesson plans for teachers, including downloadable handouts. Students can learn about food production and transport, and its impacts on the environment.

BC Ministry of Education—Environmental Concepts in the Classroom
http://www.bced.gov.bc.ca/environment_ed/
Provides a free PDF of *Environmental Learning and Experience: An Interdisciplinary Guide for Teachers* (2007).

Botanical Society of America
http://www.botany.org/

California Foundation for Agriculture in the Classroom
http://www.cfaitc.org/
Includes K-12 lesson plans.

California School Garden Network
http://www.csgn.org/
Includes resource and publication links to help with your own school garden community project.

Canadian Wildlife Federation
http://www.cwf-fcf.org/
Gives information about wildlife, habitats and other wildlife issues. Articles are available on a range of environmental and wildlife issues. The site also provides 29 lesson plans for teachers on wildlife and environmental-related topics.

The Center for Ecoliteracy
www.ecoliteracy.org

The Center for Ecoliteracy–Essay Collection
http://www.ecoliteracy.org/publications/

David Suzuki
http://www.davidsuzuki.org
Includes a wide variety of environmental information, practical steps for slowing climate change, and contests.

David Suzuki–Nature Challenge for Kids
http://www.davidsuzuki.org/kids/
Provides students with practical suggestions for environmental stewardship. Ten practical challenges and corresponding activities are presented to kids to conserve energy and be friends of the environment. Links to games, quizzes and environmental information are included as well. Sources for parents and teachers regarding environmental education can also be found on the site.

The Edible Schoolyard
http://www.edibleschoolyard.org/homepage

Education For Sustainable Development Toolkit
http://www.esdtoolkit.org/
A manual for educators and communities on how local sustainability goals can be reached through collaboration and community partnerships.

Energy Hog
http://www.energyhog.org/adult/educators.htm
An educational site for grade 2-6 teachers and students that includes both teacher and student guides, activities, handouts and quizzes.

Evergreen
http://www.evergreen.ca/en/
An extensive Canadian environmental site with tonnes of PDF links on school garden projects, including:

The Learning Grounds Guide for Schools
http://www.evergreen.ca/en/lg/lg-guide.pdf
A guide to getting school gardens/greening projects started—includes funding information.

Design Ideas for the Outdoor Classroom: Dig It, Plant It, Build It, Paint It!
http://www.evergreen.ca/en/lg/designideas.html
Ideas and techniques for creating school native plant and vegetable gardens.

Getting Started Resource List
http://www.evergreen.ca/en/lg/resources/workbox/7_Resources_Links_upda ted.pdf
Includes 11 book resources and 3 Web sites for school greening projects.

All Hands in the Dirt: A Guide to Designing and Creating Natural School Grounds
http://www.evergreen.ca/en/lg/resources/allhands/index.html
A guide to planning, designing, setting up press releases and fundraising for green schoolyard projects.

Grounds for Learning: Stories and Insights from Six Canadian School
Ground naturalization initiatives
http://www.evergreen.ca/en/lg/lg-grounds.pdf
A collection of six case studies of Canadian school greening initiatives, most
including gardens.

Stewards and Storytellers: The Greening of British Columbia School Grounds
http://www.evergreen.ca/en/lg/stewards.pdf
A collection of five case studies of British Columbian school greening initiatives, all including gardens.

Patterns through the Seasons: A Year of School Food Garden Activities
http://www.evergreen.ca/en/lg/patterns.pdf
Describes the materials, procedures and extensions related to planting a school
garden and seeing it through the four seasons. It also includes curriculum
integration information for a range of subject areas from kindergarten to grade
7.

Home and Garden Tricks and Tips
http://www.evergreen.ca/en/hg/hg-tips.html
Tips for starting a garden from scratch, maintaining a garden, cleanup and
taking care of specific flora like bulbs and evergreens.

FoodShare
www.foodshare.net

Green Teacher Magazine
http://greenteacher.com/
A quarterly print publication for educators. Each issue includes: Ideas for
rethinking education in light of environmental and global challenges, practical
articles and ready-to-use activities for various age levels from 6 to 18, and
resource reviews and evaluations of dozens of new books, kits, games and
other green resources.

LifeCycles
www.lifecyclesproject.ca

Life Lab
http://www.lifelab.org
Non-profit organization that develops curriculum and programming to study the natural world.

National Gardening Association
http://www.kidsgardening.com/
Includes information for teachers and parents about the benefits of school gardens for kids and youth. Includes project ideas and activities, classroom stories and more.

Native Plant Society of British Columbia
http://www.npsbc.org/
Includes information about gardening with native plants, education and events, and has a link to an interactive online atlas of native plants—http://www.eflora.bc.ca/

Project FEED
http://web.pdx.edu/~feed/mission.htm
Description of Project FEED, Food-Based Ecological Education Design

Setting Up and Running a School Garden: A Manual for Teachers, Parents and Communities
http://www.fao.org/docrep/009/a0218e/a0218e00.htm
A free and online comprehensive guide for all necessary aspects of getting a school garden up and running. Includes food and nutrition fact sheets and horticultural notes. More information on the guide can be found at http://www.cityfarmer.org/FAOschool.html

Seeds of Diversity
http://www.seeds.ca

Tree Canada
http://treecanada.ca/
Provides information about the benefits of trees for the community and the environment. Has resources for students and teachers as well.

The BC Working Group on Sustainability Education
http://www.walkingthetalk.bc.ca/about
Example of a social network site for individuals, organizations and government
to collaborate in order to move forward on sustainability education.

Print Resources

Greening School Grounds: Creating Habitats for Learning. Tim Grant and Gail
Littlejohn, Eds. Gabriola Island: New Society Publishers.
A practical guide for planning and executing a school garden or greening
project for any age level. Includes activities and examples, and is printed by
Green Teacher.

Last Child in the Woods: Saving Our Children from Nature-Deficit Disorder by
Richard Louv. Chapel Hill, NC: Algonquin Books of Chapel Hill, 2008.
Includes an appendix with 100 actions we can take to get children excited
about nature and the environment, including various greening initiative
projects.

Print Resources (Fiction)
- *Alison's Zinnia* by Anita Lobel. Alphabet book of plants from A through Z.
- *Jack's Garden* by Henry Cole. Rhyming story about gardening for young
  children.
- *Planting a Garden* by Lois Ehlert. Children's book about the cycle of
  planning, planting and growing a garden.
- *The Tiny Seed* by Eric Carle. Picture book that tells the story of a seed from
  its beginning to its growth into a beautiful flower.
- *How a Seed Grows* by Helene J. Jordan. Grade 1–3 book that provides an
  understanding of how seeds work.

✿

CHAPTER SEVEN

Environment, Community, Transformation: Photo Essay

Ode to Chickadee

A chickadee is tougher than you think.
One landed on my hand,
I thought it would be light,
it had weight.

I thought it would be mysterious.
Instead it was a fact
with claws
and eyes.
More like a poker player
than a poem.

Oh chickadee,
Oh chickadee,
outside silently.

Black, white and brown
flies branch to branch
all day long;
light and strong and able

to change direction
at 1/100 of a second.

(Veronica Gaylie)

Hope

Optimism cannot be commanded...but hope can be nurtured by doing good work, being open to life, and rising above our lesser selves. Hope, real hope, comes from doing the things before us that need to be done in the spirit of thankfulness and celebration, without worrying about whether we will win or lose. (Orr 2004, p.210)

Hands-On Learning

The benefits of hands-on learning are widely acknowledged among educators...Learning is a function of experience and the best education is one that is sensory-rich, emotionally engaging, and linked to the real world. (Grant & Littlejohn 2004, p.xv)

Ecology-Centred Learning

Although adaptation is the act of adapting (the decisions that are made, their evolution from initiation to implementation), adaptive capacity defines the conditions that allow (or prevent) adaptation to occur. (Cohen et al. 2006, p.18)

The University

[A] place where society can think differently, act differently, and can do so right where its citizens live. (M'Gonigle & Starke 2006, p.17)

Ecology-Centred Curriculum

• Gardens offer dynamic settings in which to integrate every discipline including science and maths, language arts, history and social studies, and art;

• Young people can experience deeper understandings of natural systems and become better stewards of the earth;

• School garden projects nurture community spirit and provide numerous opportunities to build bridges among students, school staff, families, local businesses, and community based organizations.

• Links with school gardens, school food service programmes, and local farms can ensure a fresh nutritious diet for children while teaching about sustainable food systems. (Desmond, Grieshop & Subramaniam 2004, p.26)

The Pond

Water, water, water…There is no shortage of water in the desert but exactly the right amount. (Abbey 1968, p.159)

Risk

(The Fire Pit at the Learning Garden, June 2007)

Risk taking is at the heart of teaching well. (Kohl 2004, p.32)

Sustainable Land Use

So the question arose in my mind that day: Did I choose to plant these potatoes, or did the potato make me do it?   In fact, both statements are true.
(Pollen 2002, p.xv)

Theory into **Practice**

I did not read books the first summer; I hoed beans. (Thoreau 1854, p.79)

Xeriscape

A thing is right when it tends to preserve the integrity, stability and beauty of the biotic community. It is wrong when it tends otherwise. The land ethic simply enlarges the boundaries of the community to include soils, waters, plants and animals, or collectively; the land. (Aldo Leopold 1949)

Ecoliteracy

But that afternoon I found myself wondering: What if the grammar is all wrong?...A bumblebee would probably regard himself as a subject in the garden and the bloom he's plundering for its drop of nectar as an object. (Pollen 2002, p.xiv)

Cob Development

I believe people are drawn to natural materials because they stand in opposition to the industrial appetite for disassembling the organic, for destroying the soul and spirit inherent in living materials. For a long time, machine-made perfection was the aesthetic standard. Now that may be shifting to an appreciation for the irregularity, the softness, the handmade quality and workmanship of natural building materials. (Elizabeth & Adams 2005, p.xv)

Cob is probably the oldest earth-building system, and the simplest. It requires no formwork, no ramming, no mechanized equipment, no industrial additives, and only minimal training. Cob, the material, is a combination of sand and clay, both of which can be found in many soils, with straw and water added. After thorough mixing, the stiff mud is piled onto a wall and formed by hand. Building with cob combines the most enjoyable elements of masonry, ceramic sculpture, and cooking. (Elizabeth & Adams 2005, p.117)

Participation

This participatory way of knowing, which incorporates respect and caring, is indeed a sustainable way of knowing and knowledge production. (Tripp & Muzzin 2005, p.70)

Land Ethic

Parents, educators, other adults, institutions—the culture itself—may say one thing to children about nature's gifts, but so many of our actions and messages—especially the ones we cannot hear ourselves deliver—are different. And children hear very well. (Louv 2008, p.14)

Perspective

Plant a garden. (Louv 2008, p. 364)

Community

Reinvigorating local places may just be the key to sustaining our global future. 'Think globally, act locally' was the mantra decades ago, but we never learned how. It is time to do so, and for that we must go back to school. (M'Gonigle & Starke 2006, p.14)

Wonder

My hunch is that the sense of wonder is fragile. (Orr 2004, p.24)

Where the Sidewalk Ends

There is a place where the sidewalk ends
And before the street begins...
(Shel Silverstein, from *Where the Sidewalk Ends*)

The Three Sisters

The question is whether we will embrace place at all—What happened here? What will happen here?—as a critical construct in educational theory, research and practice. (Gruenewald 2003, p.11)

Look up at the sky. Ask yourselves: Is it yes or no?...And you will see how everything changes. (Saint-Exupéry 1943, p.111)

Back to the Land

...we crave only reality. (Thoreau 1854, p.71)

A Gift, A Poem

A poem is the trace of a sojourn in the wilderness. (Leggo 1997, p.134)

CHAPTER EIGHT

❀

Conclusion: At the Heart of Teaching and Learning

I know that we can learn to love particular places. I believe that the evolution of a humane, just and durable civilization will require a great deal of love and competence and restraint that love requires for particular places. (Orr 2005c, p.103)

And now here is my secret, a very simple secret: It is only with the heart that one can see rightly; what is essential is invisible to the eye. (Antoine de Saint-Exupéry, From *The Little Prince*)

The garden has evolved into a community, its teaching a call to transformation. The garden started with tangible work; paradoxically, the intangible part of the learning process might be the garden's legacy. The garden lives. The garden teaches me it is not a garden. The plants and the boxes are 'beautiful' in one way, but the learning legacy is diversity, community, change, transformation, even love.

Without an awareness of the learning process that takes place in a garden, I can see how the place could be dismissed as a 'tool' for lessons in Ecology or Science. Except...the lessons taught in a garden are not the most important happening in a garden. What is seen in a garden is not what takes place. The garden teaches. It's easy to miss how the space itself transforms the pedagogical processes. The garden wears a disguise of flowers, pond, birds, sun and shade...and so it might be easy to dismiss the true, deep knowledge of the place. Over the different cohorts, I notice how the small, organic scale and slower pace of the garden, the gentleness of the space, always becomes part of the lesson. People come bursting into the garden with their own ideas, and they leave filled with the garden's ideas. Even in its location on an ever shrinking island of pine forest, the garden, somehow, remains hidden. The garden...makes people wonder. (From the Author's Journal, April 2008)

Building the garden made visible the larger academic institution that sur-rounds our work, and also the role of teacher education within an industrial model of learning. Tensions in the garden brought new questions: Is it enough to teach teachers how to teach? Can new teachers learn how to teach unless they learn how to learn? Wendell Berry (2002) states that a community "must change in response to its own changing needs and local circumstance, not in response to motives, powers, or fashions coming from elsewhere" (p.163). Garden-based teacher education on a college campus nurtures local learning communities in a way that involves youth, campus members and a diverse local community. The garden is also a model classroom for new teachers, where, as Berry suggests, the community and the curriculum are developed in the context of location.

The 'mandate' for a community, a garden and a classroom begins with local needs and values. When learning supports peace, community and environmental awareness, new values emerge that help learners make ecologi-cally just decisions that challenge ingrained learning patterns. In this way, a garden challenges teacher education at its very roots.

A garden might seem like a 'small' solution for much larger social and environmental issues facing the planet. In our garden I observed how outdoor work, with community and tradition at its core, interrupts an industrial model of learning. Simply put, a garden is a garden, but, in the context of education, a garden also represents an active, hands-on connection to learning that potentially shifts the emphasis from the products of education to the process of understanding. The importance of a garden lies in the fact that change occurs in process, at the grass roots, among students, at the heart of learning.

At first a garden might seem inadequate to meet the larger need to 'save' the planet. A garden seems so small, so labour intensive, with the rewards taking place in some unseen future, far from the gaze of industrial power structures. What can a garden offer students accustomed to fast technology? In fact, when a garden engages students, its very existence directly challenges what is actually *worth* knowing in 'super sized', industrial economies. Once students experience the interaction and 'small' learning in a garden, the way they approach hierarchies of learning also changes. A garden that once seemed small becomes large. The environment, once made invisible by teaching and curriculum that ignored the land, becomes visible. The magic of learning in school gardens is in revealing that which was previously hidden, in discovering something profound, in the life cycle, right outside the door. As a small-scale, local model of learning, and food economy, a garden's survival is based on tangible work, organic processes and interdependent connections instead of abstract speculation. A garden is thus sustainable because of its scale.

Alice Waters, one of the founders of the garden at Martin Luther King Middle School in Berkeley and a member of the 'slow food' movement, states:

> We act as if the seasons are of no particular consequence and the qualities of the places where the food is raised and where it is eaten are of no particular consequence. But the seasons connect us to nature. They punctuate the passage of time and they teach us about the impermanence of life—something our society hasn't wanted to look at. (Waters 2005, p.52)

What if teachers saw their students as part of the life system and honoured them not just as 'learners' but as participants in that system?

Since the very first cohort walked out to the barren site on campus and started weeding, if there has been a trend to learning in the garden, it is that education, learning and teacher training must pay attention to the processes that create it. As David Orr (2004) asks, "We have an affinity for nature. What do we do about that simple but overwhelming fact?" (p.212). A school garden invites students, and teachers, to care for the context of learning, to pay close attention to the values and 'driving force' of the places and processes that (always, invisibly) guide their lessons.

In the Learning Garden I observed how teaching and learning in nature turned students towards the idea of learning in/as process, with unique sensitivity to their surroundings. While such a transformation sounds dramatic in hindsight, at the time, we were mostly unaware of what was taking place. We stood under the sun, shovelled manure and tried to make decisions guided by Berry's ecological design intelligence, Orr's ideas around learning with earth in mind, Shiva's concept of Earth Democracy and Bowers' writing about 'revitalizing the commons'. The obvious uniqueness of the Learning Garden was that a group of teachers went outside and attempted to realize the ideals of environmental education.

The profound changes in the students' perspectives on learning were largely invisible as we discussed the best placement for the compost heap, the most responsible way to irrigate in a riparian desert, and how to build cedar wood planter boxes. I realized later how teaching in/as environment engages students in learning that becomes process, and in processes that become learning. As cohorts of student teachers began learning in the garden, even *the way* the students engaged with their own change was not something I had encountered in my seven years of training teachers, or in my previous years teaching high school students. As the students mastered the 'lesson plan' (in the first two weeks of the programme), the 'learning objectives' became conversations in the form of organic, non-linear debate. In fact, so much of the learning took place through conversation and (often) strongly expressed debate. The changes occurred so organically, so subtly and were so simple and clear in the days of the garden that by the end of the term students were usually as comfortable discussing their own transformation as they were discussing the practice and philosophy of xeriscape. The transformation by end of the term was, and is, always astonishing.

While research studies often measure school gardens on the criteria of curriculum integration, improved test scores and developing general awareness of 'place' among school students (Graham, Lane Beall, Lussier, Mclaughlin & Zidenberg-Cherr 2005), I note how our garden was suddenly at the centre of a profound shift in teacher training and research though it was only the beginning. What started as a place for exploring earth-based teaching methods soon became a change agent that dramatically shifted how students learn and how teachers teach. In the garden, I observed how the simple promotion of organic life, with sensitivity towards land and community, changed students' previously held views about learning. Perhaps the garden allows students to restore a sense of wonder, care and love for the world and for one another and perhaps because that restoration is so badly needed in the world it becomes profound.

The need for ecologically and socially inclusive teacher training, the need to nurture students' and teachers' sense of love for the land, is urgent. As Orr (2004) describes, centuries of industrial education have crushed the spirits of learners, with disastrous results: "My hunch is that the sense of wonder is fragile; once crushed it rarely blossoms again but is replaced by varying shades of cynicism and disappointment in the world" (p.24).

Clearly something is needed to nurture land, life and learning. As literary scholar Conrad Shumaker (2008) says in connection with the Navajo people's view of human interaction with the land, "after decades of 'conquering nature', we are coming to fear destruction by the forces that we have unleashed" (p.27). He describes the kind of thinking that has destroyed, literally severed, our connection with the land: "We become destroyers who know and use only parts of things, cutting them away from the whole and often causing the death of what we touch. Since our science turns everything into objects separate from the beholding mind, we find ourselves cut off from and fearful of direct contact with a nature from which our stories exclude us" (p.27).

In the garden, students learn about nature, even as they learn how to serve their social and biotic communities; the hope is that they will bring that sense of connection to their future classrooms, where they will nurture new, connected narratives around the land. Orr and Berry and others write about love in the context of nature and learning and advocate the promotion of ecoliterate values and sensitivity to one's surroundings in a way that affects how a student acquires knowledge in the first place. The promotion of ecologically just values are now vitally needed in learning; part of nurturing love for land and for one another in the context of education involves an appreciation and respect for the life cycle itself. In fact, in the garden, we learned how part of the meaning lies in the mystery, and in the wonder, in the letting alone, perhaps even in the occasional doing of nothing but letting nature teach.

The garden teaches teachers as if they were farmers; they work hard and, just as importantly, let the land do its work. In attempting to build a model outdoor learning centre called Meadowcreek, Orr (2005c) writes how, just as he tried to 'neaten' the edges of the land, the land had other ideas that included floods and droughts: "the place itself became an agent in the curriculum in ways we did not always anticipate" (p.97). Working against the force of nature, he learned how "land has a mind of its own to which we are not privy" (p.97). Orr's writings on the difficulties of building Meadowcreek provided the basis for some of his most insightful writing: "once in a lifetime we ought to give ourselves up to a particular landscape, to dwell on it, wonder about it, imagine it, touch it, listen to it, and recollect" (p.106).

In the words of an experienced farmer/educator,

> no matter how long a person has been farming, it is difficult to shake off the cultural programming that we carry with us: that farms or gardens should be made of straight rows, filled only with what we put in them; that we are somehow in control; that good farming is about technique. But as hard as we may try to mold and manipulate our farms and gardens into our own image, nature always has another idea. (Ableman 2005, p.177)

One of the tasks of the garden is to teach teachers that real learning does not always begin with subject curriculum, performance standards or lesson plans. Learning about the balance and force of nature, in the land and in ourselves, while also accepting our own human smallness, and honouring limits, is where learning begins. As we learned in the garden, such dramatic shifts in awareness took place with hardly an outward sign.

How, then, to record such 'invisible change'? Learning in a garden changes what 'counts' as knowledge. Orr (2005b) calls on Henry David Thoreau:

> He produced no particularly useful data, but he did live his subject carefully, observing Walden, its environs, and himself. In the process he revealed something of the potential lying untapped in the commonplace, in our own places, in ourselves, and the relationship between all three. (p.86)

Garden-based teacher education puts the ideals of environmental education into practice. Conceptualizing new forms of eco-centred teacher education also helps students challenge the myth of control and knowledge 'ownership' in teaching. It would be impossible for one person to build and maintain a school garden, and it would be purposeless, since land cultivation is always rooted in a process of shared knowledge.

A school garden is always, simultaneously, environment *and* community.

Teaching teachers in a Learning Garden changes what students learn, how they learn and how a teacher teaches. Teaching in nature is slow, meaningful and in fact requires a change in the scale of measurement and in the methods for recording. In our garden, in combination with community service, and in developing small-scale model economies, the natural environment teaches how to learn through a movement towards community, and in the ideal of a gift that is freely given. In the garden, students learn the importance of having something to give. It is a lesson, a gift, they will pass on to their own students. As Dewey (1897) states in *My Pedagogic Creed*, "the art of thus giving shape to human powers and adapting them to social service is the supreme art" (p.30).

As students learn to bend to the garden, and to the service of teaching, learning becomes sustainable and rooted in ideals that value mutual support, care and equality. In a garden students learn to become teachers by learning what lasts.

> Put your faith in the two inches of humus that will build under the trees every thousand years. (Berry, cited in Rowe 2006, p.49)

Learning to Balance

At the end of the term, the student who told me "I hate nature" gave a presentation. He discussed the change in himself; he now loved the Learning Garden, took weekend trips to visit community xeriscapes and showed off his new knowledge to his friends. He described to the class how garden steward-ship and community service helped him move from 'football fiend' to 'garden guru'. For students entering their extended teaching practicum in schools, the native Oregon grape, able to survive with little water, held new meaning.

All of this took place in our garden: work, community, transformation, service and, yes, love, as the students made connections to the land and learning. What takes place in teaching teachers in a garden can still, superfi-cially, seem too 'simple' for serious scholarly investigation. And I am still only beginning to experience the ways that sustainability and 'service' can occur through 'head', 'hands', and 'heart' in very real, important, ways. Still I believe that these small ways of digging and weeding together in community prompts a new model of learning and being that helps education move beyond what Canadian ecologist Stan Rowe (2006) calls "the crass promptings of utilitarian reason" (p.103) to a place where the processes of land are actually part of the lesson. Overall, I would like to see the word 'manure' more in 'serious' academic research.

Incorporating ecology into teacher training brings an awareness and sensi-tivity to surroundings and helps new teachers put empathy into practice. I want to teach my students how to connect their students to the land. I want them to look around, at the land, at each other, and I want them to learn how to respond with care and thought. In teacher education, so often, what is essential—the heart, the ability to connect to one another and to details of local place—gets washed away in a sea of coiled teacher organizers. In the garden students struggle with 'new' concepts of shared ownership and mutual success; often there is little in their previous training that has encouraged them to move beyond abstract goals.

The first year in the garden with the teachers also forced me to weed through some high-minded ideals and to get down to relevant questions such as What is important in learning? Is life important in learning? What is the role of a student teacher or a new teacher in a changing climate? What is the role of a campus garden? Eagan (1992), like Orr, suggests we do not have a choice to 'opt out': "By virtue of its legal status as titleholder...every college and university, is the obligate steward of its place on earth, regardless of whether it realizes the significance of that role" (p.66). In the garden, the students awoke to their role as land stewards, as those who act in a role of public trust. The students, now teachers, carry the same sense of ethical responsibility into their schools.

When we first began building the garden we had no idea of the profound implications of what we were involved in. We began naively, oblivious to both the institutional powers of academia and to the way our work would dramatically influence how brand new teachers would approach learning. We began by digging and planting earth with our hands, and over time learned that stewardship, *care-taking*, is an act of generosity similar to teaching itself. Stewardship became a way of thinking.

> From an ecological perspective, stewardship implies as much a way of thinking as a way of acting toward the land. It means learning about and taking care of a place, whether a natural area or urban setting, but it calls for care that is in the best interest of the place, not of the caretaker. Although definitions of stewardship vary, the human element is central: In places where people and nature coexist, human effort is needed to safeguard the integrity of the land and its inhabitants. (Eagan 1992, p.67)

Eagan's words about 'human effort', the necessity of stepping up to 'safeguard the integrity of the land' have become real for me and for students/teachers in the garden. Learning in a garden changes learning and requires care and sensitivity. A proposed model of gentleness towards the earth sounds easy. Some students enter teaching wanting to change the world, and yet the world itself is changing. The teachers are passionate and want to 'do' something to help; they want to play a role and when they think about 'good teaching' they think of someone who is hardworking, reliable and viable. They do not think of someone who arrives in class to hand out worksheets to students. However, environmental change also brings a mandate to stop harming and to stop warming. Sometimes when students stop consciously 'teaching' (or what they think of as teaching) a connection to the planet, which is silent, starts to take place. In such a context, the concept of 'viable' teaching (the root of which is *life*) must always return to the question What kind of teaching will allow life to sustain itself?

The garden has taught me how to be a teacher who leads and serves land and community. In learning how to facilitate learning in the garden, I have learned that, in the garden, nature, natural cycles, the land itself, the community itself, in fact, all aspects of the learning experience, including the teacher, teach. In a garden, teachers direct knowledge towards direct care for the planet. Instead of 'teacher'- or 'student'-centred models, an earth-centred model allows teachers to act as reflectors who use their positioning to shine light on the processes where students become communities, where local concerns are seen through a lens of global concerns. Shifting emphasis to ecology-centred learning (instead of teacher/student-centred learning *about* the environment) offers a chance to learn about environment, beyond human control.

David Orr talks about the problem of education itself, the problems of humans themselves. His imperative is that humans must begin to get away from the notion of 'all-knowingness' and start paying "full and close attention to the ecological conditions and prerequisites by which we live. That we seldom know how human actions affect ecosystems or the biosphere gives us every reason to act with informed precaution" (p.xiii).

When environmental education is slotted into single course studies or even taught as a single unit in science, perhaps it is time to immerse student teachers in a new kind of teaching and learning, one that begins with a deeper, interdisciplinary awareness of one's surroundings and will perhaps, in time, lead towards deeper respect for the earth. Developing such respect lies at the roots of ecological intelligence for student teachers, a kind of whole intelligence. Gruenewald's (2003) comments about 'a critical pedagogy of place' similarly emphasize how

> [b]eing in a situation has a spatial, geographical, contextual dimension. Reflecting on one's situation corresponds to reflecting on the space(s) one inhabits; acting on one's situation often corresponds to changing one's relationship to a place...it is this spatial dimension of situationality, and its attention to social transformation, that connects critical pedagogy with a pedagogy of place. (p.4)

Ecological justice and stewardship *as learning* begins with place, with the interaction of students and values in that place, which in turn becomes learning. Alongside their movement into the land, community and service, students gain a sense of how social and ecological transformation are intertwined, connected to that 'situationality'. Simply put, the processes of outdoor learning, in theory and practice, mirror the complexity of natural systems. The garden thus becomes both a living symbol of and a vehicle for transformation.

Learning how to balance local, global, tangible and conceptual knowledge itself provides an important core lesson and the basis for student ecological design decisions. Decisions made in gardens have immediate, visible consequences, so students learn to think hard, from a variety of ethical, ecological, human and scientific perspectives. For example, early in a garden's development, students consider the origins, present needs and future of soil, seeds and surroundings. As the process continues, students use practical knowledge within larger natural learning systems to nurture their plantings; they make decisions about water sources always within in the context of their natural surroundings. When their plantings bear fruit, and the seeds return to the soil for care by future stewards, students have already taken part in an entire life cycle. Respecting life cycles of plants within larger systems gives students firsthand understanding of the balance and details involved in sustaining life. From a place of respect and responsibility, students are then open to understanding how local ecological, social and economic systems affect global systems; students gain hands-on understanding of what it takes for small and large systems to sustain themselves.

Success in a Learning Garden involves balance. Bringing theories into action, school gardens advance environmental ideals through practice. With questions and worries over climate change increasingly a part of daily discussion, school gardens offer communities a means for peaceful involvement and critical understanding that will lead to life affirming, sustainable choices. As school gardens become widespread in North America and Europe, environmental theory can help educators include critical issues surrounding energy, food, urbanization and global economic practices that shape our daily interactions with natural systems. A basic understanding of current theories around environment can also help new educators see how their work resonates within a global community dedicated to eco-centred learning. From planting and watering seeds to developing ideals of globally sustainable economies, to the complexities of enclosure of public land and free thought, to students harvesting their own locally grown, organic vegetables at school, school gardens provide a sustainable community and outlet for ideas, actions, philosophies, traditions and a real way to care for the planet.

Starting Points

Thou saw the fields laid bare an' waste,
An' weary winter comin fast,
An' cozie here, beneath the blast,
Thou thought to dwell—
Till crash! The cruel coulter past
Out thro' thy cell.
(Robert Burns, From *On Turning Her Up in Her Nest with the Plough*)

From the first day, the story of the Learning Garden has been about the impact of local, small-scale actions on larger systems. One school garden, with sometimes just a single teacher's involvement, can produce far-reaching effects. As Shiva (2005) notes,

> small-scale responses become necessary...because large-scale structures and processes are controlled by the dominant power. The small becomes powerful in rebuilding living cultures...The large is small in terms of the range of people's alternatives. The small is large where unleashing people's energies are concerned. (p.183)

Each cohort in the garden brought its own awareness, its own action, its own legacy and its own mistakes, from which future, and present, learners could learn. After students built the Fire Pit in the garden, learning was revealed through experience and symbols and through the process of awareness of what the natural world teaches. Post-construction, the garden became an inclusive, process-based, community model of learning. The lesson of the Fire Pit was transformation; it taught teachers that transformational and ecologically and socially just learning experiences require risk taking. The legacy of the Fire Pit, dug eight feet down, now reminds students and teachers about the most basic starting point needed to learn 'about' environment: we are on the earth and we are in the earth.

While school gardens require a well-planned curriculum, strong activities and brave organizational strategies that will help create the conditions by which alternative learning settings can exist, deep learning experiences in school gardens require something more. School gardens potentially offer an incredibly varied experience and therefore require learning conditions that foster alternatives, variations and diversity. Without that 'openness' a school garden is simply an outdoor version of an indoor classroom. From our campus garden, and in visiting and reading about other school gardens, I have learned how gardens 'succeed' well in their own unique ways, on their own budgets, in their own contexts. Throughout this journey, as the garden became a learning laboratory, prompting diverse definitions of 'successful teaching' among student teachers, I also learned how all gardens are Learning Gardens. In this way, a garden teaches teachers how to learn.

I have noticed that the effects, and worth, of a school garden are difficult to explain and often even harder to prove. As described throughout this book, a garden offers a simple and profound opportunity, within the abstractions of academic learning, to include relevant, tangible work. In school gardens I find dedication, focus, relevance, hope, community, inclusivity, shared wisdom, a respect for alternatives and surprises. A garden opens up a wide field of practice and pedagogical research even as it offers a chance for meaningful, local learning. From the very beginning, the garden taught me how authentic teaching and learning requires the ability to promote, understand and move easily between hands-on experience and conceptual thought, between local and global understanding, between outsider and mainstream experience. Learning to teach in the way of the garden thus teaches students the importance of agility, the ability to bridge differences and the fact that learning and discovery is a part of the organic life cycle in both large and small ways. A garden literally brings teaching 'to life' as the lessons of human existence are mirrored, so profoundly, in organic processes in our surroundings, throughout the life (and learning) cycle.

Learning gardens show how learning about sustainability requires the conscious inclusion of values needed to sustain life in the first place. The purpose of a school garden is not only to teach *about* ecology, or to teach eco-centred lessons, but to teach *ecologically*. And, since the conditions for research in a garden are so dynamic, rooted in hands-on work and experiential learning, as researchers, the mechanisms for measurement must also remain flexible. Even if a single 'model' for school gardens could be established, any set of inflexible practices in a garden defeats the purpose of promoting connection and wonder between learners and their environments. School gardens require room for diverse practices that always arise from the students themselves. In a garden, the land and the students are also teachers.

I started the garden with the idea that we are ethically bound to teach as if the planet mattered. Given the current scientific evidence on climate change, given the fact that our socio-economic stability is closely tied to fossil fuels and that those resources are vanishing, it is vital that students, especially student teachers, gain in-depth understanding of the ways in which we depend on, and therefore must value, natural systems for our survival. We must teach students to depend on the planet in ways that are tied not only to economy but to learning and well-being, to local and global communities and to the planet. The answers do not lie in perpetuating the same unsustainable social and economic systems which caused the problems; the future will require cooperation of communities of teaching and learning as if those communities mattered within the context of the environment.

I began the garden with the idea that we can no longer teach and learn in isolation from one another and the planet. We must begin to teach and learn as if life itself depended on it. During the process of building the garden with the students, I realized the importance of Geddes' initial ideas about environmental immersion as a starting point for environmental education. Teaching about the earth's processes, and community, without an actual connection to those processes only reinforces notions of learning in abstraction and isolation.

David Orr (2004) states that the ways in which science has been traditionally researched and taught, in terms of 'discoveries' without respect for eco-centred systems of knowledge, has increased ecological destruction through scientific 'advances' in war and invasive resource exploitation. He notes that while we already have the scientific knowledge that might bring about solutions to the issues of climate change and energy consumption, the application of such knowledge involves advances in ecological literacy, in knowing how to apply such knowledge in ways that sustain the planet. In this interpretation, no matter what 'new' knowledge we have, we will not heal environmental damage with the most advanced views that ignore ecological-centred understanding.

> As important as research is, the lack of it is not the limiting factor in the conservation of biological diversity. For all that we do not know, we know without question that we are rapidly unravelling ecosystems and destabilizing the biosphere with consequences that cannot be good, which is to say that we know enough to act far more prudently, to conserve and to restore that which through carelessness we have destroyed and are destroying. Why, then, do we find it so difficult to do what is merely obvious and necessary? (Orr 2004, p.70)

An ecologically literate system of economics, education and thought involves being able to say 'no' to progress that is possible, profitable and wrong. In such an earth-centred framework, remaining whole as a planet must be both the goal and the means, where an ecologically sustainable process is a goal in itself. Orr's imperative is that we must honour the ecological knowledge that we *do not* know; as he states, the fact "that we seldom know how human actions affect ecosystems or the biosphere gives us every reason to act with informed precaution" (p.xiii).

Letting nature itself lead also invites deeper knowledge of the natural world. Such instances cannot be planned and invites students to see the limits of notions like 'student-centred learning', a view that severs students from the context of their larger learning environment. Needed is a form of teacher education that teaches students in context and that pays attention to those contexts.

My experience in our garden has taught me that the major challenge to earth-based learning is not whether students will agree to learn lessons in nature, but whether they will take part in learning processes that are so different from previous modes of learning where they found individual success. Leading students into garden-based teacher education is an entirely different space, where small ideas, learning processes and tangible knowledge matter. Eco-based teacher education, in its newness, can also lead students, initially, into some discomfort. To teach in the midst of confusion, discomfort and the unknown is not exactly a textbook model of what new teachers imagine 'good' teaching looks like. To teach student teachers in a garden is to therefore re-imagine teaching from the roots, from deeper ideals around shared responsibility where comfort is not necessarily the primary objective. In the garden, ultimately, students locate within themselves a deeper sense of care for the earth, and for the garden community, that transcends their own individual objectives. An ability to move outside of themselves, to empathize, to work hard alongside peers, to give selflessly for (an unseen) future is at the core of their transformation into teachers.

After reading an article by C.A. Bowers on the 'commons', one of the students wrote the following in the Field Notes:

> As I read page 45 outdoors, an ant crawls across the page. My immediate instinct is to brush it away, but is this book (or the space it occupies) not an example of the commons shared between all forms of life? (I do not brush it away).

What we learn, how we act. There is a direct correlation between what/how we learn and what we do/how we act. This is an obvious fact! How many teachers, though, learn to disconnect their teaching and learning from 'real life'... as if the two can be separated?

When the garden becomes part of teaching, time after time, I watch student teachers reshape their ideas about teaching and learning. Ultimately, the students are shaped by environment and learn to value the knowledge that supports environment on both a local and global scales. They learn how one species relies on another, how each system is profoundly dependent on another, and how such knowledge brings the responsibility to frame learning around new questions. They learn: "There are other plants who seem to ask of this world not riches but room" (Leopold 1949, p.103).

Notes

Chapter Two

1. Rees and Wackernagel created the 'Ecological Footprint' to "measure the total area of productive land and water required to continuously produce all resources consumed and to assimilate all wastes produced by a defined population in a specific location" (cited in Ryu and Brody 2006, p.161). The Sierra Club defines the term as "the biologically productive area required to produce the natural resources you consume expressed in global acres".
   http://www.sierraclub.org/footprint/definition.html
2. For links to school gardens and state/provincial standards, see "A Child's Garden of Standards: Linking School Gardens to California Education Standards" (2002): www.cde.ca.gov/Ls/nu/he/documents/childsgarden.pdf. Also see "The BC Ministry of Education Curriculum Supplement: Environmental Learning and Experience, an Interdisciplinary Guide for Teachers" (2007):
   http://www.bced.gov.bc.ca/environment_ed/principl.html
3. The authors describe the diversity of practice that exists in TEKW itself, and emphasize that the learning model is based on an ethical value: "Indeed, more than any other single concept, it is the notion of respect for all life forms and that land itself that characterizes [traditional] North American belief systems. Resource management was carried out through a value system that enforced practices of sustainability, expressed as respect for all life-forms" (Turner, Ignace & Ignace 2000, p.127).
4. "The commons represent both the naturals systems (water, air, soil, forests, oceans, etc.) and the cultural patterns and traditions (intergenerational knowledge ranging from growing and preparing food, medicinal practices, arts, crafts, ceremonies, etc.) that are shared without cost by all members of the community" (From the Ecojustice Dictionary: http://www.cabowers.net/CAdictmain.phpecojustice dictionary). Bowers (2006) further describes the tradition of 'the commons' as "every aspect of the human/biotic community that had not been monetized or privatized" (p. 2).
5. "The process of revitalization involves both the strengthening of local decision making in ways that ensure the continuation of these practices well as reclaiming aspects of the commons that previously had been enclosed. The latter may take the form of renewing traditional agricultural practices that do not rely upon genetically engineered seeds and

industrial chemicals, restoring wetlands, promoting community-centered arts that provide an alternative to the mass entertainment industry, the recovery of traditional knowledge and practices connected with the use of medicinal plants, the strengthening of the ties between local producers and consumers, the development of barter and local currency systems."From the Ecojustice Dictionary: http://www.cabowers.net/CAdictmain.php

6. Emory College is one example of how a campus garden can be integrated into current and future goals for campus research that also directly supports sustainability of the campus itself. The garden is used by faculties across campus and includes a healing garden of medicinal herbs and plants supported by the Department of Nursing; the campus also supports graduate research assistants in managing the garden and in incorporating the locally grown food into campus food systems. Oberlin College is another example of a campus that includes a strong, campus-wide, cross-disciplinary focus on ecology

Chapter Three

1. Climate change researchers note that "although adaptation is the act of adapting (the decisions that are made, their evolution from initiation to implementation), adaptive capacity defines the conditions that allow (or prevent) adaptation to occur" (Cohen, Neilson, Smith et al. 2006).

2. Orr (2004) describes the historical separation of learners from their environment as rooted in academic thought, surrounding a number of intellectual myths: "Historically, Francis Bacon's proposed union between knowledge and power foreshadowed the contemporary alliance between government and business, and knowledge which has wrought so much mischief. Galileo's separation of the intellect foreshadowed the dominance of the analytical mind over that part given to creativity, humour and wholeness. And in Descartes' epistemology, one finds the roots of the radical separation of self and object" (p.8).

3. Colonial appropriation of land occupied by indigenous people was in part based on the concept of terra nullius, a Latin term meaning 'land of no one'. From such a perspective, if the land was not visibly occupied, it was empty and free for the taking. "The view prevailed throughout the nineteenth century when the foundations for the Indian Act were being laid" (Report of the Royal Commission on Aboriginal Peoples 1991). Available from Cowichan Valley School District Web site: http://www.ainc-inac.gc.ca/ch/rcap/sg/sg22_e.html

Chapter Four

1. More information on the Medicine Wheel, from First Nations perspectives:
   http://www.shannonthunderbird.com/medicine_wheel_teachings.htm

2. On the historical idea of integrating aboriginal education into main stream education without acknowledgment of difference, Marie Battiste (1995) comments: "In diverse teacher-training programs, professors thought that they could give the benefit of their experience to Aboriginal schools and teachers, although they soon found that their experience was significantly different from what was needed in Aboriginal communities. The lingering question of who was to define cognitive styles embedded in Aboriginal worldviews...remained unresolved" (p.ix). Calliou (1995) adds that in the context of indigenous learners, "denial of [difference] affronts equal legitimization of everyone's voice" (p.59).

Chapter Six

1. Traditional oral account by Lois Thomas of Cornwall Island, compiled by students at Centennial College and found in "Indian Legends of Eastern Canada". Cited from Web site: Bird Clan of East Alabama. http://www.birdclan.org/threesisters.htm

2. Cited from Web site: Bird Clan of East Alabama. http://www.birdclan.org/threesisters.htm Original citation in Erney, D. (1996). Long live the Three Sisters. Organic Gardening. November: 37-40.

3. Telisa Johnston. Reprinted by permission.

4. Jenny Clark. Reprinted by permission.

5. Rochelle Lamoreaux. Reprinted by permission.

6. Davinder Bal. Reprinted by permission.

7. Frances Dixon. Reprinted by permission.

8. http://www.cobworks.com/explanation.htm. The 'Cobworks' site includes details on the use of other types of "Earth Houses" including straw bale. http://www.evergreen.ca/en/lg/patterns.pdf

9. From Hameister, R.P. (2004). Surveying Biodiversity. In Teaching Green, The Middle Years: Hands-on Learning in Grades 6-8. Tim Grant and Gail Littlejohn (Eds.). Gabriola Island, BC: New Society.

10. From Houghton, E. (2005). A Breath of Fresh Air: Celebrating Nature and School Gardens. Toronto, ON: Sumach Press.

11. From Dean, B.R. (2001). Creating a Field Guide to Your Nature Area. In Greening School Grounds. T. Grant & G. Littlejohn (Eds.). Toronto: Green Teacher.

12. Readings, J., & Taven, G. Learning Links for Elementary Classes. In Greening School Grounds. T. Grant & G. Littlejohn (Eds.). Toronto: Green Teacher.

13. Ibid.
14. Ibid
15. Ibid
16. Ibid
17. Ibid.

Bibliography

Abbey, E. (1968). *Desert solitaire: A season in the wilderness.* New York: Ballantine Books.

Ableman, M. (2005). Raising whole children is like raising good food: Beyond factory farming and factory schooling. In *Ecological literacy: Educating our children for a sustainable world.* M.K. Stone, Z. Barlow, & F. Capra (Eds.). San Francisco, CA: Sierra Club Books.

Aoki, T. (1991). Sonare and Videre: Questioning the primacy of the eye in curriculum talk. In *Reflections from the heart of educational inquiry: Understanding curriculum and teaching through the arts.* G. Willis & W.H. Schubert (Eds.). Albany: State University of New York Press.

Armstrong, J. (1999). Let us begin with courage. Centre for ecoliteracy publications page. Accessed November 2008 at
http://www.ecoliteracy.org/publications/pdf/jarmstrong_letusbegin.pdf

Armstrong, J. (2005). En'owkin: Decision-making as if sustainability mattered. In *Ecological literacy: Educating our children for a sustainable world.* M.K. Stone, Z. Barlow, & F. Capra (Eds.). San Francisco, CA: Sierra Club Books.

Arrow, C.R. (2008). The three sisters. *Bird clan of East Central Alabama Web site.* Accessed November 2008 at http://www.birdclan.org/threesisters.htm

Barlow, Z., Marcellino, S., & Stone, M.K. (2005). Leadership and the learning community. An interview with Jean Casella, principal of Mary E. Silveira School, San Rafael, California. In *Ecological literacy: Educating our children for a sustainable world.* M.K. Stone, Z. Barlow, & F. Capra (Eds.). San Francisco, CA: Sierra Club Books.

Barlow, Z., & Stone, M.K. (2005). Introduction. In *Ecological literacy: Educating our children for a sustainable world.* M.K. Stone, Z. Barlow, & F. Capra (Eds.). San Francisco, CA: Sierra Club Books.

Battiste, M. (1995). Introduction. In *First nations education in Canada: The circle unfolds*. M. Battiste & J. Barman (Eds.). Vancouver: University of British Columbia Press.

BC Ministry of Education (2007). Environmental learning and experience: An interdisciplinary guide for teachers. Available on BC Ministry of Education Web site. Accessed November 2008 at
http://www.bced.gov.bc.ca/environment_ed/

Berry, W. (2002). Sex, economy, freedom and community. In *The art of the commonplace: The agrarian essays of Wendell Berry*. Washington, DC: Shoemaker & Hoard.

Blackstock, M. (2001). Water: A first nation's spiritual and ecological perspective. B.C. *Journal of Ecosystems and Management*. 1:1-14.

Bowers, C.A. (1999). Changing the dominant cultural perspective of education. In *Ecological education in action: On weaving education, culture and the environment*. G.A. Smith & D.R. Williams (Eds.). Albany: State University of New York Press.

Bowers, C.A. (2006). *Revitalizing the commons: Cultural and educational sites of resistance and affirmation*. Lanham, MD: Lexington Books.

California Department of Education (2002). *Kids' publications, funding are newest ingredients in state's recipe for improved nutrition*. California Department of Education Newsletter. Accessed May 2008 at
http://www.cde.ca.gov/nr/ne/yr02/yr02rel35.asp

Callenbach, E. (2005). The power of words. In *Ecological literacy: Educating our children for a sustainable world*. M.K. Stone, Z. Barlow, & F. Capra (Eds.). San Francisco, CA: Sierra Club Books.

Calliou, S. (1995). Peacekeeping actions at home: A medicine wheel model for peacekeeping pedagogy. In *First nations education in Canada: The circle unfolds*. M. Battiste & J. Barman (Eds.). Vancouver: University of British Columbia Press.

Capra, F. (2005a). How nature sustains the web of life. In *Ecological literacy: Educating our children for a sustainable world*. M.K. Stone, Z. Barlow, & F. Capra (Eds.). San Francisco, CA: Sierra Club Books.

Capra, F. (2005b). Speaking nature's language: Principles for sustainability. In *Ecological literacy: Educating our children for a sustainable world*. M.K. Stone, Z. Barlow, & F. Capra (Eds.). San Francisco, CA: Sierra Club Books.

Carson, R. (1962). [2002]. *Silent spring*. New York: Houghton Mifflin.

Coffey, A. (2001). School grounds in a box: Model-making and design. In *Greening school grounds*. T. Grant & G. Littlejohn (Eds.). Toronto: Green Teacher.

Cohen, S., et al. (2006). Learning with local help: Expanding the dialogue on climate change and water management in the Okanagan region, British Columbia, Canada. *Climatic Change*. 75:3.

Comnes, L. (2005). Revolution step-by-step: On building a climate for change. An interview with Neil Smith, former principal, Martin Luther King Middle School, Berkeley, CA. In *Ecological literacy: Educating our children for a sustainable world*. M.K. Stone, Z. Barlow, & F. Capra (Eds.). San Francisco, CA: Sierra Club Books.

Consortium for School Networking (2006). Hot technologies for education: What's happening now and later? *Middle Ground*. 10:1. Accessed November 2008 at http://www.nmsa.org/Publications/MiddleGround/Articles/August2006/tabid/824/Default.aspx

Corcoran, P.B. (1999). Environmental autobiography in undergraduate educational studies. In *Ecological education in action: On weaving education, culture and the environment*. G.A. Smith & D.R. Williams (Eds.). Albany: State University of New York Press.

Curtis, D. (2001). Start with the pyramid: Real-world issues motivate students. *Edutopia* Web site. George Lucas Educational Foundation. Accessed November 2008 at http://www.edutopia.org/start-pyramid.

Desmond, D., Grieshop, J., & Subramaniam, A. (2004). *Revisiting garden based learning in basic education.* Food and Agricultural Organization (FAO) of the United Nations, Rome. Accessed November 2008 at www.fao.org/sd/erp/revisiting.pdf

Dewey, J. (1897). [2006]. My pedagogic creed. In *Critical issues in education: An anthology of readings.* Eugene F. Provenzo (Ed.). San Francisco, CA: Sage.

Dewey, J. (1902 and 1915). [1990]. *The school and society/The child and the curriculum.* Chicago: University of Chicago Press.

Dyment, J.A., & Reid, A. (2005). Breaking new ground? Reflections on greening school grounds as sites of ecological, pedagogical and social transformation. *Canadian Journal of Environmental Education.* Spring:10.

Eagan, D.J. (1992). Campus environmental stewardship. In the campus and environmental responsibility. *New Directions for Higher Education.* 1:77.

Edmundson, J. (2006). *Rethinking food: Towards a curriculum for sustainable food.* Paper presented at the annual meeting of the American Education Research Association (AERA). San Francisco, CA.

Elizabeth, L., & Adams, C. (Eds.). (2005). *Alternative construction: Contemporary natural building methods.* New York: John Wiley.

Evergreen Foundation Report (2000). *Nature nurtures: Investigating the potential of school grounds.* Vancouver, BC: Evergreen. Accessed November 2008 at http://www.evergreen.ca/en/lg/nurtures-en.pdf

Evergreen Foundation Report (2006). The outdoor classroom. The learning grounds newsletter on school grounds transformation. 18:12. Accessed November 2008 at http://www.evergreen.ca/en/lg/lg-resources.html

Graham, H., Lane Beall, D., Lussier, M. Mclaughlin, P., & Zidenberg-Cherr, S. (2005). Use of school gardens in academic instruction. *Journal of Nutrition Education and Behavior*. 37:3.

Grant, T., & Littlejohn, G. (2004). (Eds). *Teaching green, the middle years: Hands-on learning in grades 6–8*. Gabriola Island, BC: New Society Publishers.

Grant, T., & Littlejohn, G. (2001). (Eds). *Greening school grounds: Creating habitats for learning*. Gabriola Island: New Society Publishers.

Green, V.L. (2004). An exploration of school gardening and its relationship to holistic education. Urban agriculture notes. City Farmer Web site. Accessed November 2008 at http://www.cityfarmer.org/holisticeduc.html.

Gruenewald, D.A. (2003). The best of both worlds: A critical pedagogy of place. *Educational Researcher*. 32:4.

Hass, R. (2005). On watershed education. In *Ecological literacy: Educating our children for a sustainable world*. M.K. Stone, Z. Barlow, & F. Capra (Eds.). San Francisco, CA: Sierra Club Books.

HM Treasury Environment Policy (2008). The environmental policy statement of the chancellor of the exchequers departments and agencies. Accessed November 2008 at http://www.hm-treasury.gov.uk/environment_policy.htm

Holt, M. (2005). The slow school: An idea whose time has come? *Ecological literacy: Educating our children for a sustainable world*. M.K. Stone, Z. Barlow, & F. Capra (Eds.). San Francisco, CA: Sierra Club Books.

Houghton, E. (2005). *A breath of fresh air: Celebrating nature and school gardens*. Toronto: Sumach Press.

IPCC (2007). Climate change 2007: Synthesis report of the fourth assessment report of the intergovernmental panel on climate change. Accessed November 2008 at http://www.ipcc.ch/ipccreports/ar4-syr.htm

Jewell, J.R. (1908). *Agricultural education including nature study and school gardens.* Washington, DC: U.S. Bureau of Education.

Kail, A. (2006). Sustaining outdoor classrooms. *Green Teacher Magazine.* July.

Kawagley, A.O., & Barnhardt, R. (1999). Education indigenous to place: Western science meets native reality. In *Ecological education in action: On weaving education, culture and the environment.* G.A. Smith & D.R. Williams (Eds.). Albany: State University of New York Press.

Kaza, S. (1999). Liberation and compassion in environmental studies. In *Ecological education in action: On weaving education, culture and the environment.* G.A. Smith & D.R. Williams (Eds.). Albany: State University of New York Press.

Kohl, H. (2004). *I won't learn from you and other thoughts on creative maladjustment.* New York: The New Press.

Krapfel, P. (1999). Deepening children's participation through local ecological investigations. In *Ecological education in action: On weaving education, culture and the environment.* G.A. Smith & D.R. Williams (Eds.). Albany: State University of New York Press.

Kunitz, S. (2005). *The wild braid: A poet reflects on a century in the garden.* New York: W.W. Norton.

Kuokkanen, R.J. (2007). *Reshaping the university: Responsibility, indigenous epistemes and the logic of the gift.* Vancouver: University of British Columbia Press.

Leggo, C. (1997). *Teaching to wonder: Responding to poetry in the secondary classroom.* Vancouver, BC: Pacific Educational Press.

Leopold, A. (1949). *A sand county almanac.* London: Oxford University Press.

Lincoln, Y.S. (1992). Curriculum studies and the traditions of inquiry: The humanistic tradition. In *The handbook of research on curriculum.* Philip Jackson

(Ed.). New York: Macmillan Library Reference.

Louv, R. (2008). *Last child in the woods: Saving our children from nature-deficit disorder*. Chapel Hill, NC: Algonquin Books.

Marchand, B. (1984). *How food was given*. Penticton, BC: Theytus Books.

Marchand, M.J. (2007). Voices from the abyss: The role of non-pedagogical epistemologies in education. Unpublished research paper. Reprinted by permission.

Marchessault, J. (1990). Song one: The riverside. Y.M. Klein (Trans.). In *All my elations: An anthology of contemporary Canadian native fiction*. Toronto: McClelland & Steward.

Margolin, M. (2005). Indian pedagogy: A look at traditional California Indian teaching techniques. In *Ecological literacy: Educating our children for a sustainable world*. M.K. Stone, Z. Barlow, & F. Capra (Eds.). San Francisco, CA: Sierra Club Books.

Martusewicz, R. (2006). Revitalizing the commons of the African-American communities in Detroit. In *Revitalizing the commons: Cultural and educational sites of resistance and affirmation*. C.A. Bowers (Ed.). Lanham, MD: Lexington Books.

McCutcheon, N., & Swanson, A. (2001). Tips and tricks for taking kids outside. In *Greening school grounds*. T. Grant & G. Littlejohn (Eds.). Toronto: Green Teacher.

Meadows, D. (2005). Dancing with systems. In *Ecological literacy: Educating our children for a sustainable world*. M.K. Stone, Z. Barlow, & F. Capra (Eds.). San Francisco, CA: Sierra Club Books.

M'Gonigle, M., & Starke, J. (2006). *Planet U: Sustaining the world, reinventing the university*. Gabriola Island, BC: New Society.

Moyers, B. (2007). Poet Robert Bly and activist Grace Lee Boggs. *Bill Moyers Journal.* (Pod cast August 31).

Murphy, M. (2003). Education for sustainability: Findings from the evaluation study of the Edible Schoolyard. Berkeley, CA: Center of Ecoliteracy. Accessed November 2008 at http://www.ecoliteracy.org/publications/pdf/ESYFindings-DrMurphy.pdf

Mutton, M., & Smith, D. (2001). A step-by-step planning guide. In *Greening school grounds.* T. Grant & G. Littlejohn (Eds.). Toronto: Green Teacher.

National Middle School Association (2005). *This we believe: Developmentally responsive middle level schools.* Westerville, OH.

Orr, D.W. (1992). *Ecological literacy: Education and the transition to a postmodern world.* Albany, NY: State University of New York Press.

Orr, D.W. (1999). Reassembling the pieces: Ecological design and the liberal arts. In *Ecological education in action: On weaving education, culture and the environment.* G.A. Smith & D.R. Williams (Eds.). Albany: State University of New York Press.

Orr, D.W. (2004). *Earth in mind: On education, environment and the human prospect.* Washington, DC: Island Press.

Orr, D.W. (2005a). Foreword. In *Ecological literacy: Educating our children for a sustainable world.* M.K. Stone, Z. Barlow, & F. Capra (Eds.). San Francisco, CA: Sierra Club Books.

Orr, D.W. (2005b). Place and pedagogy. In *Ecological literacy: Educating our children for a sustainable world.* M.K. Stone, Z. Barlow, & F. Capra (Eds.). San Francisco, CA: Sierra Club Books.

Orr, D.W. (2005c). Recollection. In *Ecological literacy: Educating our children for a sustainable world.* M.K. Stone, Z. Barlow, & F. Capra (Eds.). San Francisco, CA: Sierra Club Books.

O'Sullivan, E. (1999). *Transformative learning: Educational vision for the 21st century*. Toronto: University of Toronto Press.

Palmer, J.A. (1998). *Environmental education in the 21st century: Theory, practice and promise*. New York: Routledge.

Pidwirny, M.J. (2002). Land use and environmental change in the Thompson-Okanagan. Royal British Columbia Museum Web site. Accessed November 2008 at http://www.livinglandscapes.bc.ca/thomp-ok/env-changes/land/ch4.html

Pollen, M. (2002). *The botany of desire: A plant's-eye view of the world*. New York: Random House.

Roberts, E., & Amidon, E. (Eds.). (1991). *Earth prayers from around the world*. San Francisco, CA: HarperCollins.

Rothstein, S.W. (1993). A short history of urban education. In *Handbook of schooling in urban America*. S.W. Rothstein (Ed.). Westport, CT: Greenwood Press.

Rowe, S. (2006). *Earth alive: Essays on ecology*. Edmonton, Canada: NeWest Press.

Ryu, H-C., & Brody, S.D. (2006). Examining the impacts of a graduate course on sustainable development using ecological footprint analysis. *International Journal of Sustainability in Higher Education*. 7:2:158–175.

Saint-Exupéry, A. de. (1943). *The little prince*. New York: Harcourt Brace Jovanovich.

Shiva, Vandana. (2005). *Earth democracy: Justice, sustainability and peace*. Cambridge: South End Press.

Shumaker, C. (2008). *Southwestern American Indian literature: In the classroom and beyond*. New York: Peter Lang.

Smith, G.A. (1999). Creating a public of environmentalists: The role of non-formal education. In *Ecological education in action: On weaving education, culture and the environment.* G.A. Smith & D.R. Williams (Eds.). Albany: SUNY Press.

Stewardship Centre for British Columbia Report (2007). Stewardship works: Supporting community-based stewardship in British Columbia. Victoria: Stewardship Centre for British Columbia. Accessed November 2008 at http://www.stewardshipworks.bc.ca/sites/ecostewardship/documents/media/31.pdf

Stone, M.K. (2005). Sustainability: A new item on the lunch menu. In *Ecological literacy: Educating our children for a sustainable world.* M.K. Stone, Z. Barlow, & F. Capra (Eds.). San Francisco, CA: Sierra Club Books.

Stone, M.K., Barlow, Z., & Capra, F. (Eds.). (2005). *Ecological literacy: Educating our children for a sustainable world.* San Francisco, CA: Sierra Club Books.

Stowell, S. (2001). Maximizing participation: Go team! In *Greening school grounds.* T. Grant & G. Littlejohn (Eds.). Toronto: Green Teacher.

Subramaniam, A. (2002). A garden based learning in basic education: A history. *University of California monograph.* 4H Centre for Youth Development: University of California, Davis.

Suzuki, D. (2002). A sense of wonder. In *Science matters.* David Suzuki foundation Web site. Accessed November 2008 at http://www.davidsuzuki.org/About_us/Dr_David_Suzuki/Article_Archives/

Syilx Nation Okanagan First Peoples Web site (2008). Accessed November 2008 at http://www.okanaganfirstpeoples.ca/elements.cfm

Thoreau, H.D. (1854). [1960]. *Walden, or, life in the woods.* New York: New American Library, Signet Classic Editions.

Thunderbird, S. (2008). Medicine wheel teachings. From Shannon Thunderbird Web site. Accessed November 2008 at http://www.shannonthunderbird.com/medicine_wheel_teachings.htm

Tracey, D. (2007). *Guerrilla gardening: A manualfesto.* Vancouver, BC: New Society.

Tripp, P., & Muzzin, L. (Eds.) (2005). *Teaching as activism: Equity meets environmentalism.* Montreal, PQ: McGill-Queens University Press.

Turner, N.J., Ignace, M.B., & Ignace, R. (2000). Traditional ecological knowledge and wisdom of aboriginal peoples in British Columbia. *Ecological Applications.* 10:5.

United Nations Educational, Scientific and Cultural Organization (UNESCO) (2005a). United Nations Decade of Education for Sustainable Development (2005–2014): International Implementation Scheme. Paris, France: UNESCO. Accessed November 2008 at http://www.unescobkk.org/fileadmin/user_upload/esd/documents/ESD_IIS .pdf

United Nations Educational, Scientific and Cultural Organization (UNESCO) (2005b). Vision & Definition of ESD. Education for Sustainable Development Web site. Accessed November 2008 at http://www.unesco.org/education/desd

United Nations Environment Programme (UNEP) Annual Report (2005). Accessed November 2008 at http://www.unep.org/AnnualReport/2005/Annual_Report12_A_Changing_ Climate.pdf

Waters, A. (2005). Fast food values and slow food values. In *Ecological literacy: Educating our children for a sustainable world.* M.K. Stone, Z. Barlow, & F. Capra (Eds.). San Francisco, CA: Sierra Club Books.

Welby, S. (2003). *Patterns through the seasons: A year of school food garden activities.* Victoria, BC: Evergreen. Accessed November 2008 at www.evergreen.ca/en/lg/patterns.pdf

Witzke, K. (2008). Conversation at Fire Pit, UBCO Learning Garden. August.